Table des matières

- 2 Introduction 4
- 3 Chapter 1: Fundamentals of AI and Machine Learning...5
  - 3.1 Overview of Artificial Intelligence 6
  - 3.2 Key Concepts in Machine Learning..7
    - 3.2.1 Supervised Learning...7
    - 3.2.2 Unsupervised Learning 7
    - 3.2.3 Neural Networks and Deep Learning 7
    - 3.2.4 Reinforcement Learning..8
    - 3.2.5 Applications in Mechanical Engineering 8
  - 3.3 AI Technologies Relevant to Mechanical Engineering..................................8
    - 3.3.1 Expert Systems 8
    - 3.3.2 Fuzzy Logic 9
    - 3.3.3 Genetic Algorithms 9
    - 3.3.4 Natural Language Processing (NLP) 9
    - 3.3.5 Applications in Mechanical Engineering 9
- 4 Chapter 2: AI in Engineering Design..11
  - 4.1 AI-Driven Design Optimization 12
  - 4.2 Generative Design and AI..13
    - 4.2.1 Key Benefits of Generative Design:..14
    - 4.2.2 Applications in Mechanical Engineering: 14
    - 4.2.3 Challenges and Future Directions:..15
  - 4.3 Case Studies: AI in Product Development 15
    - 4.3.1 Automotive Industry: AI-Driven Generative Design 15
    - 4.3.2 Aerospace Engineering: AI-Driven Design Optimization...................17
    - 4.3.3 Consumer Electronics: AI-Driven Product Development 19
- 5 Chapter 3: AI in Manufacturing Processes 22
- 6 Introduction 23
  - 6.1 . Overview of Industry 4.0 23
  - 6.2 Role of AI in Smart Manufacturing 23
  - 6.3 Case Studies and Real-World Applications..24
    - 6.3.1 Case Study 1: Siemens' Digital Factory..24
    - 6.3.2 Case Study 2: General Electric's (GE) Predix Platform.......................24

7 Chapter 4: Predictive Maintenance and AI 26
  7.1 Introduction to Predictive Maintenance 27
  7.2 1. Principles of Predictive Maintenance 27
8 Chapter 5: AI in Material Science 30
  8.1 Material Discovery using AI 31
  8.2 Enhancing Material Properties with AI 31
  8.3 Case Studies: Innovative Materials Developed with AI 31
9 Chapter 6: Optimization of Mechanical Systems 33
  9.1 AI Algorithms for System Optimization 34
  9.2 Real-Time Control and Decision Making 35
  9.3 Examples of Optimized Mechanical Systems 36
10 Chapter 7: AI and Additive Manufacturing 38
  10.1 The Role of AI in 3D Printing 39
  10.2 Design and Fabrication of Complex Structures 40
    10.2.1 AI-Enabled Design of Complex Structures 40
    10.2.2 AI-Driven Fabrication of Complex Structures 41
  10.3 Future Trends in AI-Driven Additive Manufacturing 42
11 Chapter 8: AI in Thermal and Fluid Systems 44
  11.1 AI Applications in Thermodynamics 45
  11.2 Fluid Dynamics and AI 47
    11.2.1 AI-Enhanced Fluid Dynamics Simulations 47
    11.2.2 Optimizing Fluid Flow in Pipelines 48
    11.2.3 Improving the Design of Aerodynamic Structures 48
    11.2.4 Future Directions in AI and Fluid Dynamics 49
  11.3 Case Studies in Thermal and Fluid System Optimization 50
    11.3.1 1. HVAC Systems: Optimizing Energy Efficiency and Comfort 50
    11.3.2 Industrial Fluid Systems: AI-Driven Flow Optimization 51
    11.3.3 Future Directions in AI for Thermal and Fluid System Optimization ... 52
12 Chapter 9: Challenges and Ethical Considerations 53
  12.1 Challenges in Integrating AI with Mechanical Engineering 54
  Data Quality and Availability 54
  Complexity of AI Algorithms 54
  System Integration 54

Skill Gaps and Interdisciplinary Knowledge 54

Ethics and Trust in AI Systems 55

Costs and Resources 55

12.2 Ethical Considerations in AI Applications 55

Bias and Fairness 55

Transparency and Explainability 55

Privacy and Data Protection 56

Accountability and Liability 56

Safety and Reliability 56

12.3 Future Directions and Innovations 57

Interdisciplinary Collaboration 57

Development of Regulations and Standards 57

Continuous Learning and Adaptation 58

AI-Enhanced Design and Simulation 58

Sustainable Engineering and AI 58

Human-AI Collaboration 58

13 Chapter 10: The Future of AI in Mechanical Engineering 59

13.1 Emerging Trends and Technologies 60

13.1.1 Emerging Trends and Technologies 60

13.2 The Road Ahead: AI in Engineering Education and Research 62

13.2.1 The Road Ahead: AI in Engineering Education and Research 62

14 Conclusion 65

15 Appendices 68

1 Introduction

Mechanical engineering has long been a cornerstone of technological progress, driving innovations that shape the modern world. From the steam engines of the Industrial Revolution to the sophisticated machinery of today, mechanical engineering has continually pushed the boundaries of what is possible. As we stand on the brink of a new era, the integration of Artificial Intelligence (AI) into mechanical engineering promises to revolutionize the field in ways previously unimaginable.

The convergence of AI and mechanical engineering heralds a transformative period, where intelligent systems enhance the design, analysis, and manufacturing processes. AI's ability to process vast amounts of data, recognize patterns, and make autonomous decisions is poised to unlock new levels of efficiency and precision. This fusion is not merely an incremental improvement but a paradigm shift that can redefine how engineers approach problems and create solutions.

In the design phase, AI-driven tools can optimize complex systems, leading to innovative products with improved performance and reduced development time. Machine learning algorithms can predict and mitigate potential failures, ensuring higher reliability and safety standards. In manufacturing, AI-powered automation and robotics enhance productivity, reduce human error, and enable the creation of smart factories that are responsive and adaptive to changing demands.

Moreover, the implementation of AI in mechanical engineering extends beyond traditional applications. It encompasses emerging fields such as additive manufacturing, where AI algorithms can design and produce complex geometries that were previously impossible to fabricate. In maintenance and operations, predictive maintenance powered by AI can foresee equipment failures before they occur, significantly reducing downtime and operational costs.

This book aims to explore the transformative potential of AI in mechanical engineering comprehensively. It will provide insights into the current state of AI applications, detailed case studies demonstrating successful integrations, and a forward-looking perspective on future trends. Readers will gain a deep understanding of how AI can enhance various aspects of mechanical engineering, from design and simulation to manufacturing and maintenance.

By delving into both the technical and practical aspects of AI integration, this book serves as a valuable resource for engineers, researchers, and industry professionals. It seeks to inspire innovation and foster a deeper appreciation for the possibilities that AI brings to mechanical engineering. As we embark on this journey, it is crucial to recognize that the fusion of AI and mechanical engineering is not just about technology but about reshaping the future of engineering design and manufacturing, driving progress, and creating a smarter, more efficient world.

2 Chapter 1: Fundamentals of AI and Machine Learning

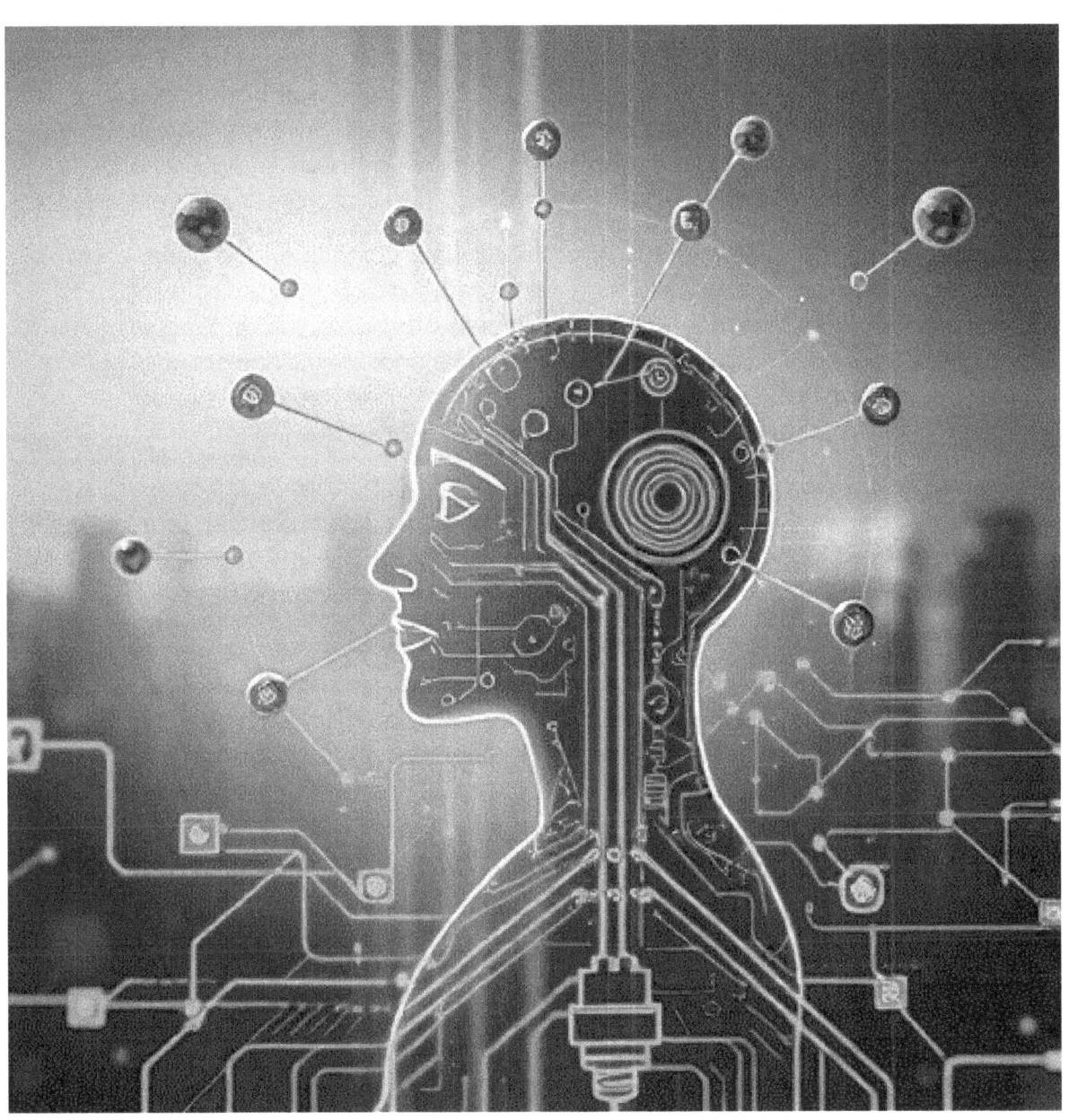

2.1 Overview of Artificial Intelligence

Artificial Intelligence (AI) is the simulation of human intelligence processes by machines, especially computer systems. These processes include learning from data, reasoning through complex problems, and self-correction to improve over time. AI encompasses a broad range of technologies and methods that enable machines to perform tasks that would typically require human intelligence. The key components of AI include knowledge representation, problem-solving, and language understanding, each contributing to the creation of systems capable of sophisticated decision-making and automation.

Knowledge representation involves encoding information about the world in a form that a computer system can utilize to solve complex tasks. This includes facts about objects, rules about their behavior, and relationships among them. Effective knowledge representation enables AI systems to understand and manipulate complex datasets, making it crucial for applications in mechanical engineering where detailed and accurate models are essential.

Problem-solving is another fundamental aspect of AI, where algorithms are designed to tackle specific challenges by searching for solutions within a defined space. Techniques such as heuristic search, optimization, and constraint satisfaction are employed to navigate through possible solutions and identify the most effective one. In mechanical engineering, problem-solving capabilities of AI can be leveraged to optimize design parameters, streamline production processes, and troubleshoot operational issues.

Language understanding, or natural language processing (NLP), allows AI systems to interpret and respond to human language. This capability is vital for creating user-friendly interfaces and enhancing human-machine interaction. In mechanical engineering, NLP can facilitate better communication between engineers and AI systems, enabling more intuitive design tools and streamlined project management.

The integration of AI into mechanical engineering opens up numerous possibilities. In design processes, AI can enhance creativity and innovation by analyzing vast amounts of data to suggest new design concepts and improvements. Machine learning algorithms can predict the performance of different designs under various conditions, reducing the need for extensive prototyping and testing. This not only accelerates the design cycle but also leads to more efficient and cost-effective solutions.

In manufacturing operations, AI can optimize production lines by monitoring equipment in real-time, predicting maintenance needs, and adjusting processes dynamically to improve efficiency and reduce waste. AI-driven automation and robotics can handle repetitive tasks with high precision, allowing human workers to focus on more complex and strategic activities. The concept of smart factories, where interconnected systems communicate and collaborate to optimize the entire production process, is becoming a reality through AI advancements.

Maintenance strategies are also significantly enhanced by AI. Predictive maintenance uses AI algorithms to analyze data from sensors embedded in machinery to foresee potential failures before they occur. This proactive approach minimizes downtime, extends the lifespan of equipment, and reduces maintenance costs. By identifying patterns and anomalies in machine performance, AI ensures that maintenance activities are performed only when necessary, thus optimizing resource allocation.

This chapter will delve deeper into the fundamental concepts of AI and machine learning, providing a solid foundation for understanding how these technologies can be applied in mechanical engineering. We will explore various AI techniques, their underlying principles, and practical applications, setting the stage for a comprehensive exploration of AI's transformative impact on the field. As we progress, you will gain insights into how AI can not only solve current engineering challenges but also pave the way for future innovations.

2.2 Key Concepts in Machine Learning

Machine Learning (ML) is a subset of AI that enables machines to learn from data and make decisions without explicit programming. By building models based on sample data, ML algorithms can predict outcomes, identify patterns, and improve their performance over time. This capability is particularly transformative in mechanical engineering, where data-driven insights can enhance design, manufacturing, and maintenance processes. ML encompasses various methods, each with unique applications and benefits.

2.2.1 Supervised Learning

Supervised learning involves training a machine learning model on a labeled dataset, where each training example is paired with an output label. The model learns to map inputs to the correct output, enabling it to make predictions or decisions based on new, unseen data. This method is widely used in mechanical engineering for tasks such as predicting material fatigue, assessing the remaining useful life of components, and optimizing design parameters. For instance, by analyzing historical data on material stress and failure points, a supervised learning model can predict when a component is likely to fail, allowing engineers to preemptively address potential issues.

2.2.2 Unsupervised Learning

In contrast to supervised learning, unsupervised learning deals with unlabeled data. The goal is to identify hidden patterns, relationships, or structures within the data. Techniques such as clustering, association, and dimensionality reduction fall under this category. In mechanical engineering, unsupervised learning can be used to cluster similar mechanical components for optimized manufacturing, identify anomalous behavior in machinery, or discover correlations between different operational parameters. For example, clustering algorithms can group parts with similar wear characteristics, enabling more efficient maintenance scheduling and inventory management.

2.2.3 Neural Networks and Deep Learning

Neural networks are computational models inspired by the human brain's structure and function. They consist of interconnected layers of nodes (neurons) that process input data to recognize complex patterns. Deep learning, a subset of neural networks, involves multiple layers (deep networks) that can handle vast amounts of data and capture intricate patterns. This approach is particularly effective for tasks such as image recognition, natural language processing, and complex system modeling.

In mechanical engineering, deep learning can be employed for quality control through image recognition, where neural networks analyze images of manufactured components to detect defects with high accuracy. Additionally, deep learning models can enhance predictive maintenance by processing sensor data from equipment to forecast potential failures. The ability of deep learning to process and learn from large, high-dimensional datasets makes it an invaluable tool for advancing automation and improving decision-making processes in engineering.

2.2.4 Reinforcement Learning

Another important aspect of machine learning is reinforcement learning, where an agent learns to make decisions by interacting with an environment. It receives feedback in the form of rewards or penalties and aims to maximize the cumulative reward over time. This approach is particularly useful for optimizing control systems and robotic operations in mechanical engineering. For example, reinforcement learning algorithms can be used to develop adaptive control strategies for robotic arms, optimizing their movements for efficiency and precision in assembly lines.

2.2.5 Applications in Mechanical Engineering

The integration of these machine learning methods into mechanical engineering offers numerous benefits. Supervised learning models can improve predictive maintenance strategies, reducing downtime and extending the lifespan of machinery. Unsupervised learning can enhance design processes by revealing hidden patterns in material properties and performance data. Neural networks and deep learning can automate quality control, ensuring high standards in manufacturing with minimal human intervention. Reinforcement learning can optimize robotic systems, enhancing automation and operational efficiency.

2.3 AI Technologies Relevant to Mechanical Engineering

The integration of AI technologies into mechanical engineering has the potential to revolutionize the field by enhancing design processes, optimizing manufacturing operations, and improving maintenance strategies. Several AI technologies are particularly relevant to mechanical engineering due to their capabilities in handling complex tasks and providing intelligent solutions. These include expert systems, fuzzy logic, genetic algorithms, and natural language processing (NLP).

2.3.1 Expert Systems

Expert systems are AI programs that emulate the decision-making abilities of a human expert. They are particularly useful in diagnostics and troubleshooting within mechanical engineering. Expert systems consist of a knowledge base containing accumulated expertise and an inference engine that applies this knowledge to solve specific problems. For instance, in a manufacturing plant, an expert system can diagnose equipment malfunctions by analyzing sensor data and historical maintenance records, offering precise recommendations for corrective actions. This technology reduces downtime and enhances the efficiency of maintenance processes, as it allows less experienced personnel to benefit from expert-level insights.

2.3.2 Fuzzy Logic

Fuzzy logic is a computational approach that deals with reasoning that is approximate rather than fixed and exact. It is highly effective in handling uncertainties and imprecisions, which are common in mechanical systems. Traditional binary logic systems operate with true or false values, but fuzzy logic allows for a continuum of truth values between 0 and 1. This flexibility makes it ideal for controlling complex mechanical systems where inputs may be vague or imprecise. For example, fuzzy logic controllers can be used in HVAC systems to maintain optimal environmental conditions by making gradual adjustments based on imprecise input data, such as "slightly warm" or "moderately humid."

2.3.3 Genetic Algorithms

Genetic algorithms are optimization techniques inspired by the process of natural selection. These algorithms are particularly useful for solving complex optimization problems in mechanical engineering. Genetic algorithms work by generating a population of potential solutions and then iteratively selecting, combining, and mutating these solutions to evolve better results over time. They are used in various applications such as optimizing the design of mechanical components, scheduling manufacturing processes, and improving the performance of engineering systems. For instance, genetic algorithms can optimize the shape and material properties of a mechanical part to achieve the best trade-off between strength and weight.

2.3.4 Natural Language Processing (NLP)

Natural Language Processing (NLP) enables machines to understand, interpret, and respond to human language. This capability is increasingly important in mechanical engineering for automated documentation, customer support, and human-machine interaction. NLP can be used to develop intelligent documentation systems that automatically generate and update technical manuals based on engineering data and user feedback. In customer support, NLP-driven chatbots can handle inquiries about machinery maintenance and troubleshooting, providing quick and accurate responses. Additionally, NLP can facilitate more intuitive interactions with AI-driven design and analysis tools, allowing engineers to input commands and queries in natural language.

2.3.5 Applications in Mechanical Engineering

These AI technologies bring significant benefits to mechanical engineering by enhancing various aspects of the field. Expert systems improve diagnostic accuracy and maintenance efficiency, reducing downtime and operational costs. Fuzzy logic offers robust control solutions in systems with inherent uncertainties, leading to better performance and reliability.

Genetic algorithms provide powerful optimization tools for design and manufacturing processes, enabling engineers to achieve superior results in less time. NLP bridges the gap between humans and machines, making advanced AI tools more accessible and easier to use..

3 Chapter 2: AI in Engineering Design

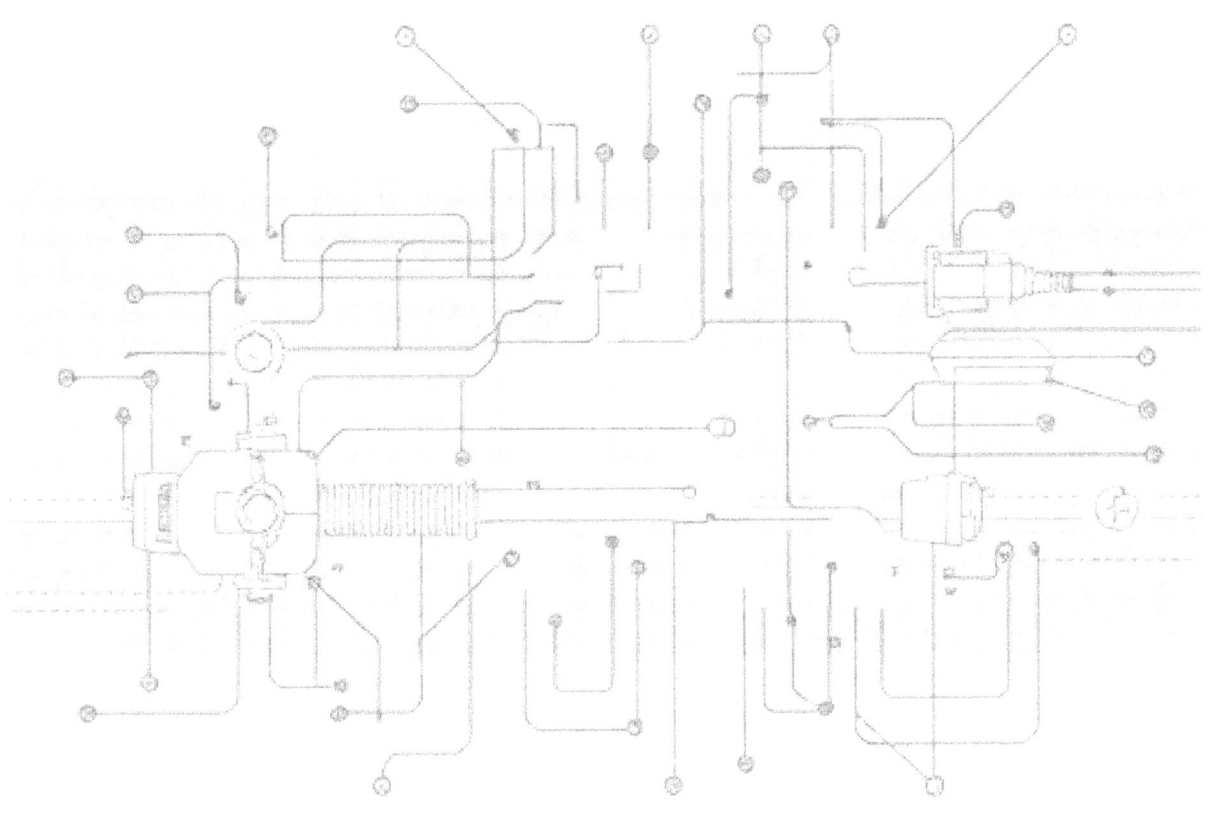

This chapter will delve deeper into the methods and applications of AI-driven design optimization, providing detailed examples and case studies that illustrate its impact on mechanical engineering. By understanding these advanced techniques, engineers and designers can harness the power of AI to create smarter, more efficient, and more innovative products.

3.1 AI-Driven Design Optimization

AI-driven design optimization is a transformative approach that leverages advanced algorithms to explore a vast space of design possibilities, ultimately identifying the most effective solutions. This method significantly enhances traditional design processes by incorporating intelligent systems capable of analyzing multiple parameters simultaneously, thereby facilitating the creation of more efficient and innovative designs.

One of the primary techniques used in AI-driven design optimization is the genetic algorithm. Inspired by the process of natural selection, genetic algorithms operate by generating a population of potential design solutions and iteratively improving them through selection, crossover, and mutation. This evolutionary approach allows for the discovery of optimal or near-optimal solutions to complex design problems that might be infeasible to solve through conventional methods. For example, in mechanical engineering, genetic algorithms can optimize the shape and material properties of a mechanical component to achieve an ideal balance between weight and strength. By simulating various design iterations, engineers can identify configurations that reduce material usage and cost while maintaining or enhancing performance and durability.

Neural networks, another powerful tool in AI-driven design optimization, excel in handling high-dimensional data and recognizing intricate patterns. Deep learning, a subset of neural networks, involves multiple layers of processing that can model complex relationships within design parameters. Neural networks can be trained on extensive datasets to predict the performance of different design configurations under various conditions. For instance, a neural network can analyze historical design data and real-time inputs to suggest modifications that improve the aerodynamic efficiency of an automotive component. This predictive capability not only speeds up the design process but also enables more innovative solutions by uncovering non-intuitive relationships between design variables.

AI-driven design optimization also incorporates techniques like reinforcement learning, where an AI agent learns to make decisions by interacting with a design environment. The agent receives feedback in the form of rewards or penalties based on the quality of the designs it generates. Over time, the agent becomes proficient in producing high-quality designs that meet predefined criteria. This approach is particularly useful in adaptive and responsive design systems where the design requirements evolve based on real-time data and changing conditions.

The application of AI-driven design optimization in mechanical engineering is vast and varied. For example, in the aerospace industry, AI can optimize the structural design of aircraft components to minimize weight and maximize fuel efficiency. In the automotive sector, AI can be used to design more efficient engines and transmission systems, leading to better performance and lower emissions. In the field of consumer electronics, AI can optimize the design of devices for improved thermal management and durability, enhancing user experience and product longevity.

Additionally, AI-driven design optimization fosters innovation by enabling engineers to explore unconventional design spaces that might be overlooked through traditional methods. Generative design, powered by AI, allows for the automatic generation of design alternatives based on specific performance criteria and constraints. Engineers can input goals such as weight reduction, material cost, and manufacturing feasibility, and the AI system generates numerous design iterations that meet these goals. This process not only accelerates the design cycle but also encourages creative and novel solutions that push the boundaries of conventional engineering practices.

Furthermore, AI-driven design optimization facilitates a more sustainable approach to engineering. By optimizing material usage and improving energy efficiency, AI helps in developing environmentally friendly designs that reduce waste and lower the carbon footprint of manufacturing processes. This aligns with the growing emphasis on sustainability in engineering and manufacturing, promoting eco-friendly innovations that benefit both industry and society.

3.2 Generative Design and AI

Generative design is an innovative and powerful approach that leverages AI algorithms to generate a wide array of design options based on specific constraints and criteria. Unlike traditional design methods, which often rely on the intuition and experience of human designers, generative design uses computational power to explore a vast design space, discovering solutions that may be unconventional yet highly efficient.

In generative design, engineers begin by inputting design goals along with essential parameters such as materials, manufacturing methods, performance requirements, and cost constraints. These inputs define the boundaries within which the AI operates. The AI then uses sophisticated algorithms to iteratively produce a multitude of design alternatives that meet the specified criteria. This process is typically driven by optimization algorithms, such as genetic algorithms, topology optimization, or machine learning models, which evaluate and refine design options based on their performance against the given constraints.

One of the most significant advantages of generative design is its ability to uncover novel solutions that a human designer might not envision. By exploring non-traditional geometries and configurations, generative design can lead to innovative structures that optimize performance, reduce material usage, and lower manufacturing costs. For instance, in the aerospace industry, generative design can create lightweight yet strong components by optimizing the internal lattice structures, leading to significant weight savings and improved fuel efficiency. Similarly, in the automotive industry, generative design can produce parts that enhance crashworthiness while minimizing weight, contributing to safer and more efficient vehicles.

3.2.1 Key Benefits of Generative Design:

- Enhanced Creativity and Innovation:

Generative design expands the creative potential of engineers by providing a diverse set of design solutions that push the boundaries of conventional thinking. This fosters innovation and leads to the development of unique products with superior performance characteristics.

- Optimized Performance:

By considering multiple performance criteria simultaneously, generative design ensures that the resulting designs are optimized for various factors such as strength, durability, weight, and thermal performance. This holistic approach leads to more robust and efficient products.

- Material and Cost Efficiency:

Generative design can significantly reduce material usage by identifying the most efficient geometries and structures. This not only lowers material costs but also contributes to more sustainable manufacturing practices by minimizing waste.

- Speed and Efficiency:

The generative design process can rapidly produce a wide range of design alternatives, accelerating the design cycle and enabling quicker decision-making. This is particularly valuable in industries where time-to-market is critical.

- Integration with Advanced Manufacturing:

Generative design is well-suited for advanced manufacturing techniques such as additive manufacturing (3D printing). The complex geometries generated by AI algorithms can be directly fabricated using 3D printing, bypassing the limitations of traditional manufacturing methods and enabling the production of highly customized parts.

3.2.2 Applications in Mechanical Engineering:

Generative design is transforming various sectors within mechanical engineering. In the architecture and construction industry, it is used to create optimized structural frameworks that balance aesthetic appeal with functional requirements. In the medical field, generative design aids in developing customized prosthetics and implants that perfectly match an individual's anatomy, improving comfort and functionality.

Moreover, generative design is playing a crucial role in the development of sustainable products. By optimizing the use of materials and enhancing energy efficiency, generative design contributes to the creation of eco-friendly products that align with global sustainability goals. For example, generative design can be used to develop lightweight components for electric vehicles, reducing their overall weight and increasing their energy efficiency.

3.2.3 Challenges and Future Directions:

While generative design offers numerous benefits, it also presents certain challenges. The complexity of the algorithms and the computational resources required can be significant, necessitating the use of powerful computers and specialized software. Additionally, integrating generative design into existing workflows may require changes in design practices and collaboration between multidisciplinary teams.

Looking ahead, the future of generative design is promising. Advances in AI and machine learning will further enhance the capabilities of generative design algorithms, enabling them to handle even more complex design problems and larger datasets. The integration of generative design with real-time data from IoT devices and digital twins will allow for the continuous optimization of products throughout their lifecycle, from design to production and maintenance

3.3 Case Studies: AI in Product Development

Real-world applications of AI in product development are transforming various industries by enhancing design efficiency, optimizing performance, and fostering innovation. In this section, we will explore how AI is being utilized in the some industry to design more aerodynamic and fuel-efficient vehicles, with a particular focus on AI-driven generative design for creating lightweight yet strong parts.

3.3.1 Automotive Industry: AI-Driven Generative Design

The automotive industry has been at the forefront of adopting AI technologies to improve vehicle design and manufacturing processes. One notable application is the use of AI-driven generative design to create lightweight yet strong automotive parts, which enhance vehicle performance and fuel efficiency.

3.3.1.1 Example: General Motors and the Seat Bracket Design

One prominent example of AI-driven generative design in the automotive industry is General Motors' (GM) collaboration with Autodesk to redesign a seat bracket. The seat bracket is a critical component that supports and secures the seat within the vehicle, requiring both strength and durability.

AI-Driven Process:

- **Defining Goals and Constraints:**

GM engineers began by inputting the design goals and constraints into the generative design software. These included requirements for strength, durability, material type, and manufacturing methods. The primary goal was to reduce the weight of the seat bracket while maintaining or enhancing its structural integrity.

- **Generative Design Algorithm:**

The generative design software, powered by AI, used these inputs to explore a vast array of design possibilities. The algorithm generated numerous design alternatives, each evaluated against the defined criteria. Through iterative processes, the software identified optimal designs that balanced weight reduction with strength and manufacturability.

- **Optimization and Selection:**

The AI-driven process resulted in the creation of multiple design options, each showcasing innovative geometries that human designers might not have envisioned. GM engineers reviewed these designs and selected the most promising one. The chosen design featured an organic, lattice-like structure that was both lightweight and strong.

- **Prototyping and Testing:**

The selected design was then prototyped using additive manufacturing (3D printing), allowing for the creation of the complex geometries generated by the AI algorithm. Rigorous testing ensued to ensure the new seat bracket met all safety and performance standards.

3.3.1.2 Results:

The AI-driven generative design process led to significant improvements in the seat bracket:

- Weight Reduction: The new design achieved a 40% reduction in weight compared to the traditional seat bracket, contributing to overall vehicle weight reduction and improved fuel efficiency.
- Increased Strength: Despite the reduction in weight, the generative design process enhanced the structural integrity of the seat bracket, ensuring it could withstand the required loads and stresses.
- Material Efficiency: The optimized design used less material, leading to cost savings in production and contributing to more sustainable manufacturing practices.
- Innovative Design: The generative design resulted in a unique, organic structure that showcased the potential of AI in creating novel and efficient engineering solutions.

3.3.1.3 Impact on the Automotive Industry:

This case study exemplifies the transformative potential of AI-driven generative design in the automotive industry. By leveraging AI technologies, automotive manufacturers can develop lighter, stronger, and more efficient components, leading to vehicles that are not only more fuel-efficient but also perform better and are safer for consumers.

Future Prospects:

The success of AI-driven generative design in creating the GM seat bracket highlights the broader implications for the automotive industry. As AI technologies continue to evolve, we can expect further advancements in vehicle design and manufacturing. Future applications may include:

- Entire Vehicle Structures: AI could optimize entire vehicle frames and body structures, maximizing safety and performance while minimizing weight.
- Electric Vehicle Components: Generative design can play a crucial role in developing more efficient battery enclosures and other components essential for electric vehicles.
- Customization and Personalization: AI can enable highly customized vehicle designs tailored to specific customer preferences and requirements, enhancing the overall driving experience.

3.3.1.4 Conclusion :

The integration of AI in product development within the automotive industry demonstrates the significant benefits of AI-driven generative design. By exploring a wide array of design possibilities and optimizing for various constraints, AI technologies are paving the way for more innovative, efficient, and sustainable automotive designs. This case study serves as a testament to the potential of AI in revolutionizing product development and setting new standards for engineering excellence.

3.3.2 Aerospace Engineering: AI-Driven Design Optimization

The aerospace industry is characterized by its stringent requirements for safety, performance, and efficiency. AI-driven design optimization plays a crucial role in meeting these requirements by enabling the creation of aircraft components that are both lightweight and capable of withstanding extreme conditions encountered in space and high-altitude flight.

3.3.2.1 Example: NASA and the Design Optimization of Spacecraft Components

NASA has been at the forefront of utilizing AI to enhance the design and performance of spacecraft and satellite components. One notable example is the application of AI-driven generative design to optimize the structural components of spacecraft, aiming to reduce weight while ensuring durability and resilience.

AI-Driven Process:

- **Defining Goals and Constraints:**

NASA engineers begin by defining the design goals and constraints for spacecraft components. These include parameters such as load-bearing capacity, thermal resistance, material properties, and manufacturing feasibility. The primary goal is to minimize the weight of the components while maximizing their strength and ability to withstand harsh space conditions, such as extreme temperatures and radiation.

- **Generative Design Algorithm:**

The generative design software, powered by AI, uses these inputs to explore a wide array of design possibilities. The algorithm iteratively generates and evaluates numerous design alternatives, optimizing for factors such as structural integrity, thermal management, and material efficiency. This process involves advanced computational techniques, including topology optimization and machine learning.

- **Optimization and Selection:**

The AI-driven process produces multiple design options, each presenting innovative geometries that traditional design methods might not consider. NASA engineers review these designs and select the most promising one. The selected design often features complex, organic shapes that provide optimal performance characteristics while being lightweight.

- **Prototyping and Testing:**

The chosen design is prototyped using advanced manufacturing techniques, such as additive manufacturing (3D printing), which allows for the precise creation of complex geometries. The prototype undergoes rigorous testing to ensure it meets all performance and safety standards required for space missions.

3.3.2.2 Results:
The AI-driven generative design process led to significant improvements in the design of spacecraft components:

- Weight Reduction: The optimized designs achieved substantial weight reductions compared to traditional components, which is critical for space missions where every gram counts in terms of cost and feasibility.
- Enhanced Strength: Despite the reduced weight, the generative design process enhanced the structural strength of the components, ensuring they could withstand the extreme conditions of space travel.

- Thermal Efficiency: The AI-optimized designs improved thermal management, helping components maintain optimal operating temperatures in the harsh environment of space.
- Material Efficiency: The designs utilized materials more efficiently, resulting in cost savings and more sustainable manufacturing practices.

3.3.2.3 Impact on Aerospace Engineering:
This case study exemplifies the transformative potential of AI-driven generative design in aerospace engineering. By leveraging AI technologies, NASA and other aerospace organizations can develop components that are lighter, stronger, and more efficient, thereby enhancing the overall performance and reliability of spacecraft and satellites.

3.3.2.4 Future Prospects:
The success of AI-driven generative design in optimizing spacecraft components highlights its broader implications for aerospace engineering. Future applications may include:

- Full Spacecraft Optimization: AI could optimize the entire structure of spacecraft, including fuselage, wings, and internal systems, to achieve unprecedented levels of efficiency and performance.
- Satellite Design: AI-driven design can enhance the development of satellite structures, improving their durability, performance, and lifespan in orbit.
- Exploration Missions: AI can assist in designing specialized components for exploration missions to other planets and celestial bodies, ensuring robustness and adaptability to unknown environments.

3.3.2.5 Conclusion:
The integration of AI in product development within aerospace engineering demonstrates the significant benefits of AI-driven generative design. By exploring a wide array of design possibilities and optimizing for various constraints, AI technologies are paving the way for more innovative, efficient, and resilient aerospace designs. This case study serves as a testament to the potential of AI in revolutionizing product development and setting new standards for engineering excellence.

3.3.3 Consumer Electronics: AI-Driven Product Development
Consumer electronics companies are constantly striving to innovate and deliver products that meet the evolving needs and preferences of consumers. AI-driven product development plays a crucial role in this process by enabling the creation of devices with improved functionality, usability, and efficiency.

3.3.3.1 Example: AI Optimization of Electronic Components Layout

One notable application of AI in consumer electronics is the optimization of the layout of electronic components within devices such as smartphones, tablets, and laptops. By leveraging AI algorithms, companies can design products that not only perform better but also offer enhanced user experiences and longevity.

AI-Driven Process:

- **Defining Goals and Constraints:**

Consumer electronics companies begin by defining the design goals and constraints for their products. These may include factors such as performance targets, thermal management requirements, battery life, and form factor considerations. The primary goal is to optimize the layout of electronic components to maximize functionality and usability while minimizing heat generation and energy consumption.

- **AI-Based Layout Optimization:**

AI algorithms are employed to analyze various layout configurations of electronic components within the device. These algorithms use techniques such as machine learning and optimization to explore a wide range of design possibilities and identify the most optimal layout. The AI evaluates each layout based on criteria such as component proximity, signal interference, and thermal dissipation.

- **Iterative Design Process:**

The AI-driven design process is iterative, with the algorithms continuously refining and improving the layout based on feedback and performance metrics. Through multiple iterations, the AI converges on a layout that achieves the desired balance between functionality, efficiency, and user experience.

- **Validation and Testing:**

The optimized layout is then prototyped and subjected to rigorous testing to validate its performance under real-world conditions. This includes tests for thermal management, signal integrity, battery life, and overall usability. Any necessary adjustments are made based on the test results before finalizing the design.

3.3.3.2 Results:
The AI-driven optimization of electronic components layout yields several key benefits:

- Improved Functionality: By optimizing the layout of electronic components, consumer electronics companies can enhance the overall functionality and performance of their products. Components are strategically positioned to minimize signal interference, optimize power distribution, and maximize processing efficiency.
- Enhanced User Experience: Devices with optimized layouts offer a more intuitive and user-friendly experience. Components are arranged to improve ergonomics, reduce clutter, and streamline user interactions. This results in products that are easier to use and more enjoyable for consumers.
- Reduced Heat Generation: One of the primary challenges in consumer electronics design is managing heat generated by electronic components. AI optimization helps mitigate this issue by strategically positioning components to improve airflow and

thermal dissipation. This not only prevents overheating but also extends the lifespan of the device.
- Energy Efficiency: Optimized layouts can also contribute to improved energy efficiency by minimizing power losses and optimizing power delivery. This leads to longer battery life and reduced energy consumption, enhancing the sustainability of consumer electronics products.

3.3.3.3 Impact on Consumer Electronics Industry:

The integration of AI-driven layout optimization into product development processes has significant implications for the consumer electronics industry. Companies can deliver products that not only meet but exceed consumer expectations in terms of performance, usability, and reliability. By leveraging AI technologies, consumer electronics companies can maintain a competitive edge in a rapidly evolving market.

3.3.3.4 Future Prospects:

As AI technologies continue to advance, we can expect further innovations in consumer electronics design and development. Future applications may include:

- Personalized Product Design: AI algorithms can analyze user preferences and behavior data to tailor product designs to individual consumers, offering personalized experiences and features.
- Sustainable Design Practices: AI optimization can help minimize the environmental impact of consumer electronics by reducing energy consumption, extending product lifespans, and promoting recyclability.
- Advanced Form Factors: AI-driven design can unlock new possibilities for device form factors, enabling innovative and unconventional designs that push the boundaries of traditional product design.

3.3.3.5 Conclusion:

The integration of AI-driven layout optimization into consumer electronics product development represents a paradigm shift in the industry. By leveraging AI technologies, companies can create products that deliver superior performance, enhanced user experiences, and increased sustainability. This case study highlights the transformative potential of AI in shaping the future of consumer electronics and setting new standards for innovation and excellence.

4 Chapter 3: AI in Manufacturing Processes

5 Introduction

Smart manufacturing leverages AI to create more flexible and efficient production processes. Industry 4.0 refers to the fourth industrial revolution, characterized by the integration of digital technologies, including AI, into manufacturing. This includes the use of AI for real-time monitoring, predictive maintenance, and autonomous decision-making.

5.1. Overview of Industry 4.0

Industry 4.0, or the Fourth Industrial Revolution, is a transformation that brings together advanced manufacturing techniques with IoT (Internet of Things), AI, big data, and cloud computing. This revolution enhances the manufacturing sector by introducing smart systems that communicate, analyze, and use information to drive further intelligent actions. The primary pillars of Industry 4.0 include:

- **Cyber-Physical Systems (CPS):** Integrating physical manufacturing processes with digital systems, where machines are equipped with sensors and actuators that connect to the digital world.
- **IoT:** Connecting devices and systems to collect and exchange data, enabling real-time decision-making.
- **Big Data and Analytics:** Using advanced analytics to process vast amounts of data generated by IoT devices and CPS.
- **Cloud Computing:** Providing scalable computing resources and data storage to handle the large volumes of data generated.

5.2 Role of AI in Smart Manufacturing

AI plays a pivotal role in enabling the core principles of Industry 4.0 by enhancing automation, improving efficiency, and enabling more informed decision-making. Key applications of AI in smart manufacturing include:

Real-Time Monitoring and Quality Control: AI algorithms analyze data from sensors in real-time to detect anomalies and ensure products meet quality standards. This reduces waste and increases the consistency of production.

Predictive Maintenance: Machine learning models predict equipment failures before they occur, scheduling maintenance proactively to minimize downtime and reduce costs. This involves analyzing historical data and identifying patterns that precede equipment failures.

Supply Chain Optimization: AI optimizes supply chain operations by forecasting demand, managing inventory levels, and selecting suppliers. This results in cost reductions and improved delivery times.

Autonomous Decision-Making: AI systems can make decisions autonomously, adjusting production schedules, reconfiguring manufacturing lines, and responding to changing market demands without human intervention.

Robotics and Automation: AI-powered robots perform complex tasks with precision and adaptability. These robots can work alongside humans or independently to increase productivity and flexibility in manufacturing processes.

5.3 Case Studies and Real-World Applications

5.3.1 Case Study 1: Siemens' Digital Factory

Siemens has implemented AI and Industry 4.0 principles in their Amberg Electronics Plant, transforming it into a digital factory. The plant utilizes AI for predictive maintenance, real-time monitoring, and autonomous decision-making. The result is a highly efficient production process with a defect rate of nearly zero.

5.3.2 Case Study 2: General Electric's (GE) Predix Platform

GE's Predix platform harnesses AI to provide industrial data analytics. It enables predictive maintenance and operational optimization across various manufacturing sectors. By analyzing data from sensors on machinery, Predix can predict failures and suggest maintenance schedules, reducing downtime and operational costs.

5.3.2.1 Challenges and Future Directions

Despite the advantages, integrating AI in manufacturing presents several challenges:

Data Security and Privacy: As more devices connect to the internet, ensuring the security of data becomes critical. Manufacturers must implement robust cybersecurity measures.

Interoperability: Ensuring different systems and devices can communicate seamlessly is a significant challenge. Developing standards for interoperability is essential for the success of Industry 4.0.

Skilled Workforce: There is a need for a workforce skilled in AI, data analytics, and digital technologies. Training and development programs are crucial to bridge this skills gap.

Initial Investment: The cost of implementing AI and digital technologies can be high. Manufacturers must evaluate the long-term benefits and ROI to justify these investments.

5.3.2.2 Future Directions

Edge Computing: Moving data processing closer to the source of data (machines and sensors) to reduce latency and improve real-time decision-making.

Advanced Robotics: Developing more sophisticated robots with enhanced AI capabilities to handle complex and varied manufacturing tasks.

Sustainable Manufacturing: Leveraging AI to create more sustainable manufacturing processes, reducing waste and energy consumption.

5.3.2.3 Conclusion

AI in manufacturing processes is a cornerstone of Industry 4.0, driving significant advancements in efficiency, flexibility, and productivity. As technologies continue to evolve, smart manufacturing will become even more intelligent and integrated, leading to a future where factories are highly autonomous, adaptive, and sustainable.

6 Chapter 4: Predictive Maintenance and AI

6.1 Introduction to Predictive Maintenance

Predictive maintenance (PdM) is a proactive maintenance strategy that leverages AI and machine learning (ML) to predict when equipment is likely to fail. By analyzing data from sensors and historical records, PdM identifies patterns and anomalies indicative of potential issues, allowing for maintenance activities to be scheduled accordingly. This approach significantly reduces downtime, extends machinery lifespan, and lowers maintenance costs by preventing unexpected breakdowns and optimizing the use of maintenance resources.

6.2 1. Principles of Predictive Maintenance

Predictive maintenance is based on the following key principles:

- **Data Collection:** Sensors collect data on various parameters such as temperature, vibration, noise, and pressure from machinery.
- **Data Processing:** The collected data is processed and cleaned to ensure accuracy and consistency.
- **Data Analysis:** Advanced algorithms analyze the processed data to identify patterns and predict potential failures.
- **Decision Making:** Maintenance activities are scheduled based on the insights derived from data analysis, ensuring timely intervention.

6.2.1.1 Machine Learning for Fault Diagnosis

Machine learning plays a crucial role in predictive maintenance by enabling fault diagnosis through the analysis of sensor data. Key techniques used in ML for fault diagnosis include:

Anomaly Detection: Identifying deviations from normal operating conditions. Anomalies may indicate potential faults that require attention. Techniques such as clustering and neural networks are often used for anomaly detection.

Regression Analysis: Predicting the remaining useful life (RUL) of machinery components by analyzing trends in sensor data over time. Regression models help forecast when a component is likely to fail.

Classification Models: Classifying the type of fault based on patterns observed in the data. Models like decision trees, support vector machines (SVM), and deep learning are used to categorize different fault types.

6.2.1.2 Example Algorithms:

Random Forest: Used for both classification and regression tasks, random forests can analyze complex datasets and provide robust predictions.

Neural Networks: Deep learning models can handle large volumes of data and learn intricate patterns, making them suitable for fault detection and diagnosis.

Support Vector Machines (SVM): Effective for classification tasks, SVMs can distinguish between normal and faulty conditions.

6.2.1.33. Real-World Applications and Case Studies
Predictive maintenance has been successfully implemented across various industries, showcasing its versatility and effectiveness.

6.2.1.4 Manufacturing Plants:
In manufacturing, AI-driven predictive maintenance systems continuously monitor machinery health. For instance, sensors on motors and pumps collect vibration and temperature data, which is analyzed to predict bearing failures or overheating. This allows maintenance teams to address issues before they cause unplanned downtimes, thus ensuring a smoother production process and reducing costs associated with emergency repairs.

6.2.1.5 Case Study: Siemens
Siemens implemented predictive maintenance in their production lines, using AI to monitor and analyze data from machine tools. By predicting failures before they occur, Siemens has significantly reduced downtime and maintenance costs while improving overall equipment efficiency.

6.2.1.6 Aerospace Industry:
Airlines utilize AI to monitor aircraft components such as engines, landing gear, and avionics systems. Predictive models analyze data from flight operations and maintenance logs to predict when parts will require maintenance. This proactive approach enhances safety, reliability, and operational efficiency, reducing the risk of in-flight failures and optimizing maintenance schedules.

6.2.1.7 Case Study: Delta Airlines
Delta Airlines employs predictive maintenance using AI to monitor and analyze engine performance data. By predicting potential engine failures, Delta has improved aircraft reliability and reduced maintenance-related delays, leading to cost savings and increased passenger satisfaction.

6.2.1.8 Energy Sector:
In the energy sector, power plants and oil rigs use predictive maintenance to prevent costly equipment failures. AI models analyze sensor data from turbines, generators, and drilling equipment to detect anomalies and predict failures. This helps in optimizing maintenance schedules, reducing operational costs, and ensuring uninterrupted energy production.

6.2.1.9 Case Study: BP (British Petroleum)
BP has implemented AI-driven predictive maintenance in their oil rigs, using machine learning to analyze data from drilling equipment. By predicting equipment failures and scheduling timely maintenance, BP has reduced downtime and maintenance costs, while enhancing operational safety and efficiency.

6.2.1.10   Challenges and Future Directions :
While predictive maintenance offers numerous benefits, it also presents several challenges :

Data Quality and Integration: Ensuring high-quality, consistent data from various sources can be challenging. Integrating data from legacy systems with modern AI solutions requires robust data management practices.

Algorithm Accuracy: Developing accurate predictive models requires extensive training data and continuous model refinement. Ensuring the models remain accurate over time involves regular updates and retraining.

Implementation Costs: Initial investment in sensor technology, data infrastructure, and AI solutions can be substantial. Organizations must assess the long-term ROI to justify these costs.

Workforce Skills: A skilled workforce is essential to develop, implement, and maintain predictive maintenance systems. Training and upskilling employees in AI and data analytics is crucial.

6.2.1.11   Future Directions :

Edge Computing: Processing data closer to the source (e.g., on the machinery itself) to reduce latency and enhance real-time predictive capabilities.

Enhanced AI Models: Developing more sophisticated AI models that can handle diverse and complex datasets, improving fault detection and prediction accuracy.

Integration with IoT: Expanding the integration of IoT devices to gather more comprehensive data, providing deeper insights into equipment health and performance.

Sustainable Maintenance: Leveraging AI to develop maintenance strategies that not only optimize equipment performance but also minimize environmental impact.

6.2.1.12   Conclusion :

Predictive maintenance powered by AI and machine learning revolutionizes the way industries approach equipment maintenance. By predicting failures before they occur, organizations can significantly reduce downtime, extend the lifespan of machinery, and lower maintenance costs. As AI technology continues to advance, predictive maintenance will become even more accurate and accessible, driving further efficiencies across various sectors.

7 Chapter 5: AI in Material Science

7.1 Material Discovery using AI
Material Discovery Using AI

The discovery of new materials has historically been a time-consuming and resource-intensive process, often involving trial-and-error experimentation. However, with the advent of artificial intelligence (AI), this process has been significantly accelerated. AI-driven approaches, particularly through machine learning, enable scientists to predict the properties and performance of potential materials before they are even synthesized in the lab. Machine learning models can sift through enormous amounts of historical data on materials to identify trends, relationships, and correlations that may not be obvious to human researchers. These models can propose new materials based on desired properties, like enhanced strength, durability, or thermal conductivity, tailored for specific applications. For example, AI can suggest novel lightweight composites for aerospace applications where material strength and low weight are critical or recommend heat-resistant alloys designed for the extreme conditions of power plants. The ability to discover and design materials more quickly opens up possibilities for technological advancement across a broad range of industries, from renewable energy to healthcare.

7.2 Enhancing Material Properties with AI

Beyond discovering new materials, AI plays a pivotal role in improving the properties of existing materials. By analyzing a material's composition and processing parameters, AI models can suggest adjustments that lead to superior performance. For instance, in metallurgy, AI can identify optimal combinations of alloying elements that result in metals with higher strength, increased durability, or better resistance to corrosion. The use of AI in polymer science can streamline the curing processes of advanced materials, ensuring they achieve their desired properties with minimal waste or defects. For complex material systems like composites, AI can optimize layer configurations and resin systems to achieve the best mechanical and thermal performance. Moreover, these optimizations are achieved with greater speed and efficiency than traditional methods, allowing manufacturers to develop high-performance materials tailored to their specific needs in a fraction of the time.

7.3 Case Studies: Innovative Materials Developed with AI

AI-driven research in material science has already yielded a range of cutting-edge materials, demonstrating the transformative potential of this approach. Some notable examples include:

Graphene-based Materials: Graphene, a single layer of carbon atoms arranged in a hexagonal lattice, is renowned for its exceptional electrical, thermal, and mechanical properties. However, producing high-quality graphene at scale has been a challenge. AI has been instrumental in optimizing the production processes of graphene, enabling its use in a wide array of high-performance applications, such as flexible electronics, energy storage systems, and high-strength composites. By fine-tuning production parameters, AI models have reduced defects and increased efficiency, bringing graphene closer to mainstream industrial use.

High-entropy Alloys (HEAs): HEAs are composed of five or more metallic elements, each in significant proportions. These materials possess unique properties, such as exceptional strength, resistance to wear, and corrosion at high temperatures, making them ideal for extreme environments like aerospace or nuclear applications. AI has been employed to explore the vast composition space of HEAs, identifying combinations that provide superior performance. Through predictive modeling, AI accelerates the discovery of HEAs, enabling scientists to focus on the most promising candidates for further study.

Smart Materials: Materials that can change their properties in response to external stimuli, such as temperature, pressure, or electrical signals, are known as smart materials. AI is being used to design such materials for applications in sensors, actuators, and other adaptive technologies. By predicting how materials will react to specific conditions, AI aids in the development of materials that can adapt in real time, making them invaluable for emerging fields like wearable technology, robotics, and medical devices. These smart materials can self-heal, change color, or even alter their mechanical properties depending on the environment, offering revolutionary possibilities in design and function.

8 Chapter 6: Optimization of Mechanical Systems

8.1 AI Algorithms for System Optimization

In the realm of mechanical systems, optimizing for factors such as cost, performance, durability, and efficiency can often present complex, multi-dimensional challenges. Traditional optimization methods, while effective in certain scenarios, may struggle to handle the complexity and vast solution space that many modern engineering problems entail. AI algorithms, such as genetic algorithms (GAs) and neural networks, have emerged as powerful tools for system optimization, capable of addressing these challenges with greater accuracy and efficiency.

Genetic Algorithms (GAs) mimic the process of natural selection, iteratively refining a population of potential solutions based on their performance against predefined criteria. In mechanical system optimization, GAs can explore a wide range of design parameters, including material choices, component shapes, and system configurations. By prioritizing and "breeding" the most successful solutions in each generation, these algorithms gradually converge on an optimal or near-optimal solution. This approach is particularly valuable in multi-objective optimization problems, where engineers must balance competing factors like cost, performance, and durability. For example, in automotive engineering, GAs might be used to find the optimal design for a car chassis that minimizes weight (for better fuel efficiency) while maximizing structural integrity and crash resistance.

Neural Networks (NNs), particularly deep learning models, are another AI tool used in system optimization. By analyzing large datasets of past designs, performance metrics, and operational conditions, NNs can learn to predict the behavior of mechanical systems under various conditions. These models are adept at handling non-linear relationships between variables, making them highly effective in optimizing systems where interactions between components are complex and difficult to model with traditional methods. For instance, in robotics, neural networks can optimize the configuration and movement patterns of robotic arms to achieve the highest precision with the lowest energy consumption. Similarly, in aerodynamics, neural networks can assist in optimizing the shape of aircraft components to reduce drag and improve fuel efficiency.

The integration of AI algorithms into mechanical system design enables engineers to evaluate far more design possibilities than could be feasibly explored through manual methods. Moreover, these algorithms can identify solutions that strike the best possible balance between multiple objectives, taking into account trade-offs and constraints that may be overlooked in conventional optimization processes. As a result, AI-driven optimization is reshaping fields such as automotive engineering, aerospace, energy systems, and manufacturing by producing more efficient, cost-effective, and high-performing mechanical systems.

8.2 Real-Time Control and Decision Making

AI's ability to process vast amounts of data in real time has revolutionized control systems in various mechanical and industrial applications. By integrating AI into these systems, engineers and operators can achieve a level of automation and precision that would be impossible through traditional control methods. AI-powered systems can continuously monitor, analyze, and adjust operations based on changing conditions, ensuring optimal performance while responding dynamically to unforeseen circumstances.

One prominent example is in HVAC (Heating, Ventilation, and Air Conditioning) systems. These systems are essential for maintaining comfort in buildings, but they can also be significant energy consumers. Traditional HVAC systems often rely on static settings or basic sensors to adjust temperature and airflow, but AI introduces a higher level of control by analyzing multiple variables in real time. AI algorithms, such as reinforcement learning or predictive models, can optimize HVAC operations by balancing factors like indoor temperature, humidity, occupancy levels, and outdoor weather conditions. By doing so, AI can maintain occupant comfort while simultaneously minimizing energy consumption, leading to significant cost savings and reduced environmental impact. For example, an AI system might predict when a room will be occupied and adjust temperatures in advance, or reduce heating and cooling in areas of the building that are unoccupied.

In manufacturing environments, AI's role in real-time control and decision-making is equally transformative. AI-driven systems can monitor equipment performance, product quality, and process variables to make continuous adjustments that maximize efficiency and product consistency. For instance, in automated assembly lines, AI can control the speed, force, and precision of robotic arms, ensuring that products are assembled to exact specifications while minimizing defects. AI systems can also predict equipment maintenance needs based on real-time data, allowing for preventive maintenance that reduces downtime and prolongs the life of machinery. This real-time decision-making capability improves operational efficiency, reduces waste, and enhances overall product quality.

AI also enables advanced predictive control systems, where decisions are made not just based on current conditions, but also by predicting future states. For example, in wind turbine farms, AI can optimize the alignment and operation of turbines based on wind speed forecasts, ensuring maximum energy capture while preventing mechanical strain on the turbines. In complex energy grids, AI-based control systems can dynamically balance supply and demand, ensuring stable and efficient energy distribution.

The integration of AI for real-time control and decision-making is becoming indispensable in industries where precision, efficiency, and adaptability are paramount. As AI algorithms continue to evolve, their ability to handle increasingly complex systems will further enhance automation, reduce operational costs, and improve the overall performance of mechanical and industrial systems.

8.3 Examples of Optimized Mechanical Systems

The application of AI to mechanical systems has led to groundbreaking improvements in efficiency, performance, and sustainability across various sectors. From transportation to energy and industrial manufacturing, AI-driven optimization is pushing the boundaries of what these systems can achieve. Below are some key examples of optimized mechanical systems powered by AI:

Automated Vehicles : Automated vehicles, including self-driving cars and drones, rely heavily on AI to optimize their operations. AI systems in these vehicles continuously analyze vast amounts of data from sensors, cameras, and GPS to make real-time decisions that improve safety and efficiency. For instance, AI-powered route planning algorithms optimize driving patterns by selecting the most fuel-efficient routes, avoiding traffic congestion, and even adjusting speeds to reduce energy consumption. Additionally, AI optimizes the way vehicles handle complex driving environments, enhancing safety by predicting and responding to potential hazards, such as pedestrians, other vehicles, or changes in road conditions. In commercial logistics, self-driving trucks and delivery drones benefit from AI-driven optimization to improve fuel economy, reduce delivery times, and lower operational costs, making transportation more sustainable and efficient.

Energy Systems: AI has revolutionized renewable energy systems, such as wind turbines and solar panels, by optimizing their operation to maximize energy production and minimize costs. In wind farms, AI algorithms analyze wind speed, direction, and turbine efficiency in real time to adjust the pitch of the blades and orientation of the turbines for optimal energy capture. Similarly, solar energy systems benefit from AI-based forecasting tools that predict weather patterns and optimize the positioning of solar panels to maximize sunlight absorption throughout the day. AI also plays a critical role in energy storage systems, ensuring that excess energy generated by renewables is stored efficiently and released when demand is high. By integrating AI into the management of energy grids, utilities can better balance supply and demand, reduce energy waste, and lower operational costs, all while increasing the overall reliability of the grid.

Industrial Processes: In industrial settings, AI-driven optimization has transformed processes such as chemical production, manufacturing, and material processing. AI systems can continuously monitor variables such as temperature, pressure, flow rates, and chemical concentrations to ensure that processes are operating at peak efficiency. In chemical production, for example, AI algorithms can minimize waste and energy consumption while maximizing yield by adjusting process parameters in real time. In material processing industries, AI can optimize the heating, cooling, and forming of materials to ensure consistent quality and reduce defects. Furthermore, AI can predict equipment failure or maintenance needs before they lead to costly downtime, enabling preventive maintenance schedules that reduce operational disruptions. The use of AI in industrial processes has led to significant cost reductions, improved product quality, and enhanced environmental sustainability by minimizing resource consumption and waste production.

9 Chapter 7: AI and Additive Manufacturing

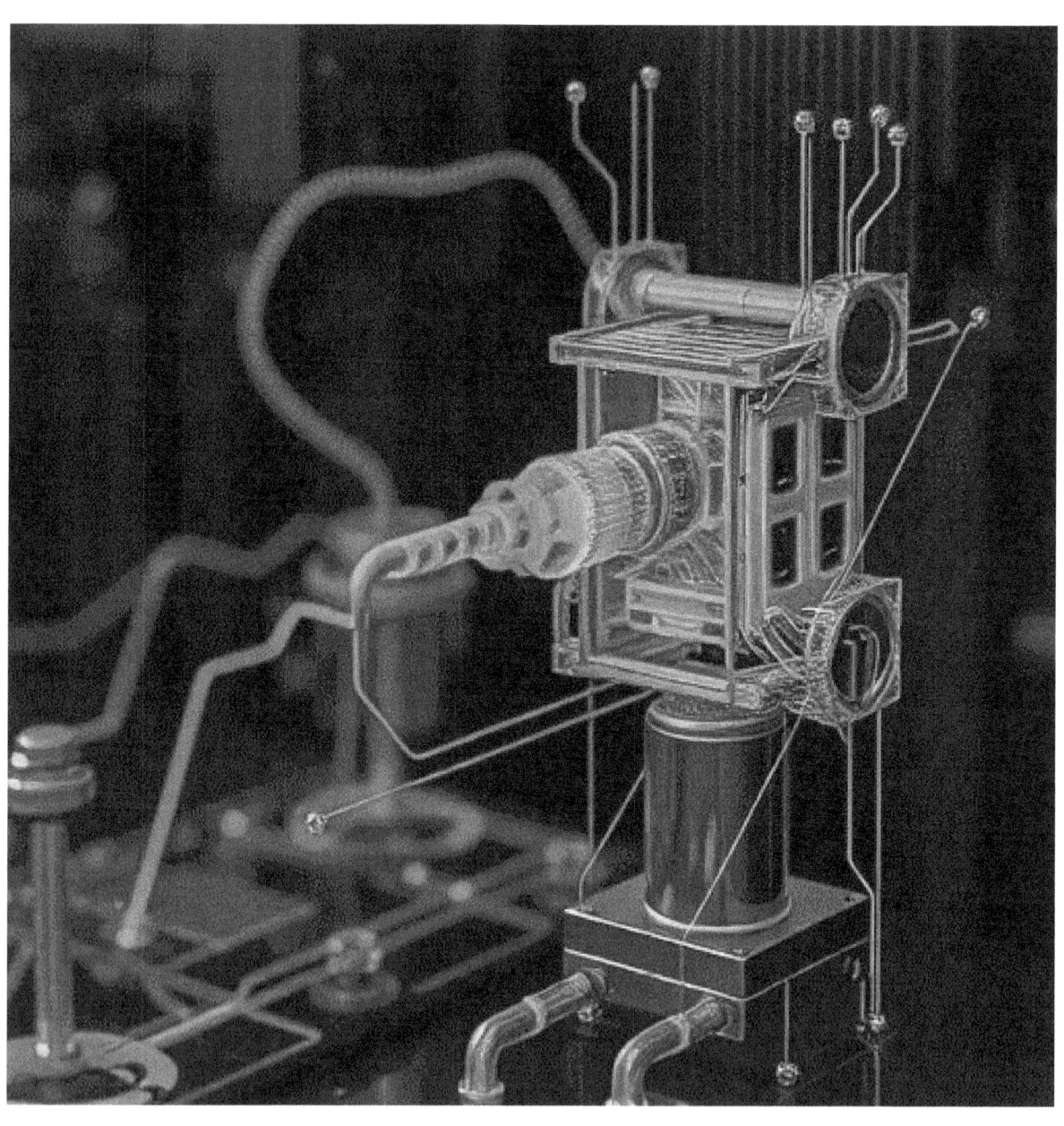

Additive manufacturing, commonly known as 3D printing, has transformed industries by enabling the production of complex geometries and customized components with unprecedented precision and flexibility. The integration of artificial intelligence (AI) into 3D printing processes has further elevated this technology, making it smarter, more efficient, and capable of producing higher-quality results. AI enhances every stage of the additive manufacturing workflow, from design optimization to real-time monitoring and quality control.

9.1 The Role of AI in 3D Printing

AI plays a crucial role in additive manufacturing by optimizing both the design and production processes, ensuring that parts are printed with the highest possible quality and efficiency.

Design Optimization: One of AI's most significant contributions to 3D printing is in the optimization of designs before production even begins. AI algorithms can analyze part geometries, material properties, and intended use cases to suggest design modifications that improve strength, reduce material usage, and minimize print time. For example, AI can generate lightweight lattice structures within a component, which provide the necessary strength while using less material, making the final product lighter and more cost-effective. This is particularly valuable in industries like aerospace, where weight reduction is critical. AI-driven generative design algorithms can also explore multiple design alternatives simultaneously, offering engineers innovative solutions that may not have been discovered using traditional design methods.

Process Optimization: AI is instrumental in improving the actual 3D printing process. AI models can predict potential issues in the printing process, such as warping, cracking, or incomplete fusion of layers, which could compromise the quality of the final product. By analyzing data from sensors embedded in 3D printers, AI systems can detect anomalies in real-time, such as temperature fluctuations or material flow inconsistencies, and make adjustments on the fly. This ability to predict and correct errors during the printing process ensures that parts are manufactured with high precision, reducing the need for post-processing or reprints.

Material Optimization: In additive manufacturing, material properties and behavior during the printing process are critical to the quality of the final product. AI can analyze how different materials perform under various printing conditions, enabling better selection of materials for specific applications. For example, in metal 3D printing, AI can optimize laser parameters like power and speed to ensure that metals are fused correctly, minimizing defects like porosity or weak points in the material. By learning from previous prints, AI can continuously refine the printing process to produce more reliable and consistent results across different materials, whether metals, polymers, or composites.

Real-time Monitoring and Quality Control: AI's ability to monitor and analyze data in real-time is particularly valuable in ensuring the quality of 3D-printed parts. By processing data from sensors, cameras, and thermal imaging systems, AI can identify deviations from the ideal printing conditions and intervene before a defect occurs. This real-time quality control system significantly reduces material waste and increases production efficiency. Additionally, AI-based predictive maintenance systems can monitor the health of 3D printers, scheduling maintenance only when necessary, reducing downtime, and prolonging the life of the machines.

The integration of AI into additive manufacturing is transforming the industry, making 3D printing faster, more reliable, and more cost-effective. From optimizing part designs to ensuring flawless production processes, AI empowers engineers to push the boundaries of what is possible with additive manufacturing, allowing for the creation of highly customized, high-performance parts across industries such as healthcare, aerospace, and automotive.

9.2 Design and Fabrication of Complex Structures

The advent of AI in conjunction with advanced manufacturing technologies, such as additive manufacturing (3D printing), has opened up new possibilities for the design and fabrication of complex structures that were previously unattainable using traditional methods. These structures, often characterized by intricate geometries, organic shapes, and tailored material distributions, offer enhanced performance, functionality, and customization across a range of industries. AI-driven design tools allow engineers and designers to explore innovative solutions that optimize key factors like strength, weight, and material efficiency, while AI-powered fabrication techniques ensure precision and scalability in production.

9.2.1 AI-Enabled Design of Complex Structures

One of the most significant contributions of AI in modern design is its ability to optimize complex structures, such as lattice and topologically optimized designs, which offer a high strength-to-weight ratio. These structures are particularly useful in industries where minimizing weight is critical, without compromising on structural integrity, such as in aerospace or medical implants.

Lattice Structures: AI can generate lattice structures, which are frameworks of repeating geometric patterns that create a highly efficient and lightweight architecture. By using machine learning algorithms, AI can analyze the specific load conditions a part will experience and tailor the lattice geometry accordingly. This ensures that material is distributed only where it's needed, significantly reducing weight while maintaining strength and durability. In aerospace, for instance, AI-optimized lattice structures can be used to create lightweight components like turbine blades or structural supports, resulting in fuel savings and improved performance. Similarly, in the medical field, AI-optimized lattices can be used to create custom-fitted implants, such as bone scaffolds or orthopedic devices, that match the patient's anatomy while offering the necessary mechanical support.

Generative Design: AI-driven generative design tools enable engineers to input design goals (e.g., strength, weight, cost) and constraints (e.g., material type, manufacturing process) into an algorithm, which then generates thousands of design options that meet those criteria. These algorithms can explore a broader solution space than traditional design methods, often coming up with novel shapes and structures that human designers may not have considered. This process is particularly valuable in fields like aerospace, automotive, and construction, where weight reduction and material efficiency are critical to performance. Generative design allows for the creation of parts that are not only more efficient but also more aesthetically pleasing and optimized for the manufacturing process, whether that be 3D printing, casting, or milling.

9.2.2 AI-Driven Fabrication of Complex Structures

Fabricating the intricate structures designed by AI is made possible by advanced manufacturing techniques like 3D printing, which can produce geometries that would be impossible with conventional subtractive methods such as machining or molding. AI enhances this fabrication process by ensuring that complex structures are built with the required precision and quality.

Topological Optimization: AI's role in topological optimization goes beyond design. During fabrication, AI algorithms can control the precise deposition of materials, ensuring that the complex structures are built layer by layer with exacting accuracy. For instance, in metal additive manufacturing, AI can monitor the fusion process, adjusting laser power and speed to avoid defects like porosity or cracking. This ensures that even highly complex geometries, such as internal lattice structures or hollow forms with intricate supports, are manufactured without compromising structural integrity.

Custom Medical Implants: One of the most promising applications of AI in the fabrication of complex structures is in custom medical implants. AI can analyze medical imaging data, such as CT or MRI scans, to generate custom implant designs that perfectly match a patient's anatomy. These designs can include complex, porous structures that promote osseointegration, where the bone grows into the implant, leading to better long-term outcomes. Additive manufacturing allows these customized designs to be fabricated with precision, ensuring that the implant fits the patient's body perfectly, reducing recovery time, and improving the implant's durability.

Aerospace Applications: In the aerospace industry, AI-driven fabrication is used to create lightweight components that optimize fuel efficiency and performance. Complex structures such as lightweight brackets, heat exchangers, and engine components can be manufactured using 3D printing methods. AI ensures that these parts meet stringent safety and performance requirements, while also minimizing material use. The ability to create such intricate, optimized parts enables aerospace manufacturers to push the boundaries of design, reducing weight and improving the overall efficiency of aircraft and spacecraft.

AI's role in the design and fabrication of complex structures is reshaping industries that rely on high-performance, lightweight, and customized components. By optimizing lattice structures, leveraging generative design, and enhancing the precision of advanced manufacturing processes, AI allows engineers to create innovative structures that are both functional and efficient, paving the way for advancements in aerospace, medical, and many other fields.

9.3 Future Trends in AI-Driven Additive Manufacturing

As AI continues to evolve and integrate with additive manufacturing technologies, the potential for innovation in 3D printing is expanding rapidly. The future of AI-driven additive manufacturing promises to deliver smarter machines, new advanced materials, and a shift towards mass customization, making manufacturing more efficient, flexible, and accessible across industries. These advancements will enable manufacturers to achieve unprecedented levels of precision, adaptability, and personalization in product design and production.

AI-Integrated 3D Printers

One of the most exciting future trends is the development of **AI-integrated 3D printers**, where AI systems are embedded directly into the printers to enable real-time monitoring and adjustment during the printing process. These next-generation machines will use advanced AI algorithms to continuously monitor key variables such as temperature, material flow, and layer adhesion, ensuring that every print is of the highest quality. By analyzing sensor data and making on-the-fly adjustments, AI will be able to correct deviations in real-time, such as shifting temperatures or inconsistencies in material deposition, preventing defects like warping or delamination. This not only reduces waste and the need for reprinting but also enhances the reliability of the process for high-stakes applications, such as aerospace or medical implants, where precision and quality are paramount.

In addition, AI-integrated 3D printers will be equipped with **self-learning capabilities**. They will be able to learn from previous prints, improving their performance with each iteration. For instance, a 3D printer could automatically adjust its parameters for different materials or geometries based on the data collected from prior print jobs, resulting in faster, more efficient production cycles. As these printers become more autonomous, they will require less manual intervention, reducing the skill barrier for operators and making additive manufacturing more accessible to a wider range of industries and users.

Advanced Materials

The development of **new materials specifically designed for additive manufacturing** is another significant trend driven by AI. Traditionally, many materials used in 3D printing were adapted from other manufacturing processes, which limited their performance and applicability. However, AI is helping to accelerate the discovery of novel materials tailored for the unique requirements of additive manufacturing, such as materials with specific mechanical properties, thermal resistance, or bio-compatibility.

AI algorithms can analyze vast amounts of data on material properties, performance, and behavior during the printing process, enabling researchers to discover **composite materials** or **metal alloys** that can be optimized for strength, flexibility, or lightweight properties. In industries like aerospace and automotive, AI-driven material design will lead to components that are not only lighter and more durable but also more cost-effective to produce. In the medical field, AI can help develop **bio-compatible materials** for implants or tissue engineering, allowing for the creation of custom, patient-specific medical devices.

Furthermore, AI can also optimize the **material formulation** for multi-material 3D printing, where different materials are combined in a single print to achieve unique properties. This capability is particularly relevant in the development of **smart materials** that can change shape, color, or mechanical properties in response to external stimuli, such as temperature or light. AI will play a critical role in fine-tuning the behavior of these materials, opening up new possibilities for innovative applications, from responsive medical implants to adaptive structures in construction.

Mass Customization

The future of manufacturing will be defined by **mass customization**, where AI enables the production of personalized products on a large scale. In traditional manufacturing, customization typically requires costly and time-consuming retooling or redesign. However, AI-driven additive manufacturing will allow manufacturers to efficiently produce customized products without sacrificing economies of scale.

AI-powered design tools, such as **generative design** and **parametric modeling**, allow users to create personalized designs that are optimized for individual requirements. These tools can take input from customers—such as specific measurements, aesthetic preferences, or functional requirements—and automatically generate custom designs that are ready for 3D printing. For example, in the footwear industry, AI can be used to analyze a customer's foot shape and biomechanics, creating a perfectly tailored shoe that offers optimal comfort and support. Similarly, in the medical sector, AI can enable the mass production of **customized prosthetics, orthotics, or dental implants** designed to match the unique anatomy of each patient.

As AI advances, the ability to produce **one-off custom products** at scale will transform industries such as fashion, consumer electronics, and healthcare. Consumers will have access to personalized products that meet their exact needs and preferences, while manufacturers can reduce waste by producing items only when ordered, rather than relying on mass production and inventory.

The future of AI-driven additive manufacturing is filled with exciting possibilities. AI-integrated printers, advanced materials, and mass customization are just a few of the trends that will reshape the landscape of manufacturing. These advancements will not only increase the efficiency and flexibility of production but also enable more sustainable, innovative, and personalized solutions across various industries, from aerospace and healthcare to consumer goods and beyond.

10 Chapter 8: AI in Thermal and Fluid Systems

10.1 AI Applications in Thermodynamics

Thermodynamics, the study of heat transfer, energy conversion, and the behavior of systems under various thermal conditions, plays a crucial role in numerous industries. From power generation and refrigeration to chemical processing and aerospace, the efficient management of thermal systems is essential for optimizing performance and reducing energy consumption. With the integration of AI, thermodynamics is undergoing a significant transformation, as AI-driven solutions provide new ways to optimize heat transfer processes, improve energy efficiency, and design advanced thermal systems.

Optimizing Heat Transfer Processes

One of the most impactful applications of AI in thermodynamics is in optimizing heat transfer processes. Heat transfer is fundamental to many industrial systems, such as power plants, HVAC systems, and manufacturing processes. AI algorithms can analyze and predict heat transfer behavior under various operating conditions, allowing for more efficient design and operation of these systems.

- **AI-Optimized Heat Exchangers:** Heat exchangers are widely used in industries to transfer heat between two or more fluids, and optimizing their performance is critical for improving energy efficiency. Traditional heat exchanger design relies on empirical data and simplified mathematical models, which may not capture the complexities of fluid dynamics and heat transfer interactions in real-world applications. AI offers a more advanced approach by using machine learning algorithms to analyze large datasets on fluid properties, flow rates, temperatures, and pressure drops. By continuously learning from this data, AI can optimize the design of heat exchangers for maximum efficiency, suggesting configurations that reduce thermal resistance and minimize energy loss. This results in better-performing heat exchangers that save energy and reduce operating costs.

- **Real-Time Control of Heat Transfer Systems:** In addition to design optimization, AI can be applied to the **real-time control** of heat transfer systems. For example, AI can monitor the performance of industrial cooling systems or heating processes in real-time, adjusting flow rates, temperatures, and pressure levels to optimize heat transfer efficiency. This is particularly valuable in large-scale systems like power plants or data centers, where maintaining optimal heat dissipation is critical for system reliability and performance. AI algorithms can predict when system parameters deviate from optimal conditions and make automatic adjustments to prevent overheating or energy loss, improving overall system efficiency.

Improving Energy Efficiency

AI is also being used to enhance energy efficiency in thermal systems by optimizing the conversion of heat into usable energy and minimizing energy waste. This is especially important in sectors like power generation, where thermodynamic efficiency directly affects overall energy output and environmental impact.

- **AI in Power Generation:** In thermal power plants, which rely on converting heat energy into mechanical work and electricity, AI can optimize processes like combustion, steam generation, and turbine performance. By analyzing real-time data from sensors placed throughout the power plant, AI can detect inefficiencies or potential bottlenecks in the energy conversion process and suggest adjustments to operating conditions. For example, AI can optimize fuel-air mixtures in combustion systems to ensure complete combustion, reducing fuel consumption and emissions. Similarly, AI can monitor the performance of steam turbines, identifying optimal operating temperatures and pressures to maximize energy extraction from steam while minimizing energy losses.

- **Energy Recovery Systems:** AI can also improve the efficiency of energy recovery systems, such as **waste heat recovery** units, which capture excess heat from industrial processes and convert it into useful energy. AI algorithms can optimize the design and operation of these systems by analyzing the temperature profiles and flow conditions of waste streams, suggesting ways to maximize heat recovery without disrupting the primary process. This helps industries reduce their overall energy consumption and lower operating costs, while also contributing to sustainability efforts by reducing waste heat emissions.

Designing Advanced Thermal Systems

AI is playing a pivotal role in the design of **advanced thermal systems**, allowing engineers to develop innovative solutions that push the boundaries of thermodynamic performance. Whether it's in the development of high-efficiency engines, next-generation cooling systems, or novel thermal storage technologies, AI-driven design tools are enabling the creation of more sophisticated and optimized thermal systems.

- **AI-Driven Thermal Storage Systems:** Thermal energy storage systems, which store excess heat or cold for later use, are becoming increasingly important in renewable energy and industrial applications. AI can optimize the design of these systems by modeling the heat storage capacity, thermal conductivity, and insulation properties of various materials. For instance, AI can help develop phase-change materials (PCMs) that absorb or release heat at specific temperatures, allowing for more efficient energy storage and retrieval. This is particularly valuable in renewable energy applications, where thermal storage can balance the intermittent nature of solar or wind power by storing excess energy during peak production times and releasing it when demand is high.

- **Advanced Cooling Systems:** AI is also transforming the design of cooling systems, particularly in high-performance industries like aerospace and electronics. In these fields, managing heat dissipation is critical to maintaining system performance and longevity. AI can optimize cooling system designs by predicting heat generation and distribution in complex environments, such as within jet engines or computer chips. AI algorithms can suggest advanced cooling techniques, such as liquid cooling, microchannel heat sinks, or even **nanofluids**—fluids enhanced with nanoparticles that improve heat transfer properties.

Future Directions

As AI continues to advance, its applications in thermodynamics will expand into even more innovative areas. For example, AI could be used to design **self-regulating thermal systems**, where materials or components adjust their thermal properties autonomously in response to changing conditions. In aerospace, AI could help develop ultra-efficient propulsion systems that operate at extreme temperatures, while in renewable energy, AI could optimize the performance of thermal energy storage systems for large-scale grid integration.

AI's integration into thermodynamics is revolutionizing heat transfer processes, energy efficiency, and the design of advanced thermal systems. By leveraging AI's ability to process complex data and predict system behavior, engineers can optimize thermal systems to achieve higher performance, lower energy consumption, and improved sustainability across a variety of industries.

10.2 Fluid Dynamics and AI

Fluid dynamics, the study of how fluids (liquids and gases) move and interact with their surroundings, is a critical field in engineering, physics, and environmental sciences. It plays a pivotal role in industries ranging from aerospace and automotive to energy and environmental management. Traditional fluid dynamics analysis, which relies on solving complex mathematical equations like the Navier-Stokes equations, often requires significant computational power and time. However, the integration of AI is revolutionizing this field by enabling faster, more accurate simulations and predictions of fluid behavior, thus improving system performance, design, and efficiency.

10.2.1 AI-Enhanced Fluid Dynamics Simulations

AI-driven simulations offer a significant improvement over traditional computational fluid dynamics (CFD) methods. Machine learning models, particularly neural networks, can be trained on vast amounts of simulation and experimental data to predict fluid behavior in real-time. These AI-enhanced simulations are not only faster but also allow engineers to explore more complex fluid interactions that would be too computationally expensive using traditional methods.

- **Predicting Turbulence:** One of the most challenging aspects of fluid dynamics is the accurate prediction of **turbulence**, the chaotic and unpredictable nature of fluid flow. Turbulence is a critical factor in many applications, from aircraft stability to weather forecasting. Traditional methods of turbulence modeling, such as Reynolds-averaged Navier-Stokes (RANS) equations or large eddy simulations (LES), are computationally intensive and often provide approximate results. AI is changing this by offering data-driven models that can predict turbulence more accurately and efficiently. Machine learning algorithms can be trained on high-fidelity turbulence data, learning complex patterns in the flow and predicting turbulent behavior in real-time. This allows engineers to optimize designs for improved aerodynamics, reduced drag, and enhanced fuel efficiency in vehicles or airplanes.

- **Reduced-Order Models for Real-Time Simulations:** In many industrial applications, running high-resolution CFD simulations is impractical due to time constraints. AI offers an alternative through **reduced-order models** (ROMs), which simplify the simulation process by reducing the number of variables while maintaining accuracy. AI can generate ROMs from high-fidelity CFD data, allowing for near-instantaneous simulations of fluid flow in complex systems. These models are particularly useful in scenarios where quick decisions are needed, such as real-time monitoring of fluid systems in power plants or optimizing fuel flow in engines.

10.2.2 Optimizing Fluid Flow in Pipelines

AI is also enhancing the design and operation of fluid transport systems, such as pipelines and ducts used in industries like oil and gas, water treatment, and HVAC systems. In these systems, optimizing fluid flow is essential for reducing energy consumption, minimizing pressure losses, and ensuring efficient transport.

- **Flow Optimization in Pipelines:** In large pipeline networks, which transport liquids and gases over long distances, maintaining optimal flow conditions is a challenge due to friction, turbulence, and changing demand. AI can be applied to monitor and optimize fluid flow in real-time, ensuring that the system operates at peak efficiency. For example, AI models can analyze data from sensors placed along the pipeline to predict pressure drops, flow rates, and potential blockages. By identifying inefficiencies or potential issues early, AI can suggest operational adjustments, such as changing pump speeds or valve positions, to minimize energy consumption and maintain steady flow. In industries like oil and gas, where pipeline networks are extensive and energy costs are high, AI-driven flow optimization can result in significant cost savings and reduced environmental impact.

- **Leak Detection and Prevention:** AI can also enhance the detection of leaks in fluid transport systems. Traditional leak detection methods often rely on pressure monitoring, which can be slow and inaccurate. AI, however, can analyze real-time data from multiple sensors and use pattern recognition algorithms to detect even the smallest anomalies in flow rates or pressure. This allows for faster and more accurate identification of leaks, reducing downtime and minimizing environmental damage in industries such as oil and gas or water distribution.

10.2.3 Improving the Design of Aerodynamic Structures

In the field of aerodynamics, AI is being used to optimize the design of structures such as aircraft, cars, and wind turbines, where the interaction between fluid flow and solid surfaces plays a crucial role in performance. By leveraging AI, engineers can create more efficient and innovative designs that reduce drag, improve stability, and maximize energy output.

- **AI-Driven Aerodynamic Design:** Traditionally, the design of aerodynamic structures involved iterative testing using wind tunnels or CFD simulations, which can be time-consuming and costly. AI, particularly through **generative design** and **reinforcement learning**, can accelerate this process. In generative design, AI algorithms take initial design parameters, such as shape constraints or desired performance outcomes, and generate numerous design alternatives that optimize fluid flow around the structure. These designs are then refined through simulation, with the AI learning from each iteration. This allows engineers to explore novel and unconventional shapes that minimize drag, improve fuel efficiency, or increase lift in the case of aircraft.

- **Optimizing Wind Turbines:** Wind turbines are a prime example of how AI can improve aerodynamic performance. The efficiency of a wind turbine depends on the design of its blades, which must capture wind energy effectively while minimizing drag. AI can optimize blade geometry by analyzing wind patterns and flow behavior, suggesting designs that maximize energy capture in different wind conditions. Additionally, AI can monitor turbine performance in real-time, adjusting the pitch and orientation of the blades to ensure optimal energy production. This is particularly valuable in offshore wind farms, where environmental conditions are constantly changing, and maximizing energy output is critical to the viability of the project.

- **Automotive Aerodynamics:** In the automotive industry, reducing aerodynamic drag is key to improving fuel efficiency and vehicle performance. AI is being used to optimize the shape and surface features of vehicles, from the overall body design to finer details like side mirrors, spoilers, and undercarriage components. By simulating airflow over the vehicle and using AI to adjust design parameters, engineers can reduce drag, improve handling, and increase fuel efficiency. AI-driven aerodynamic optimization is particularly important in the development of electric vehicles (EVs), where range is a key factor, and minimizing energy losses is crucial for extending battery life.

10.2.4 Future Directions in AI and Fluid Dynamics

The future of AI in fluid dynamics is likely to see even more advanced applications, such as **AI-assisted multi-phase flow simulations** and **AI-driven fluid-structure interaction models**. These complex phenomena, which involve the interaction of multiple fluid phases (e.g., liquid and gas) or the interaction between fluids and solid structures, are critical in fields like offshore engineering, chemical processing, and environmental studies. AI's ability to process vast amounts of data and predict complex interactions will enable more accurate and efficient simulations of these phenomena.

Additionally, the combination of **AI and high-performance computing** (HPC) will allow for more detailed and faster fluid dynamics simulations at a scale previously unimaginable, further revolutionizing industries that rely on fluid dynamics.

AI's impact on fluid dynamics is transforming how engineers and scientists approach the design, optimization, and control of fluid systems. From predicting turbulence and optimizing pipeline flow to improving aerodynamic structures, AI is unlocking new possibilities in fluid dynamics that offer greater efficiency, precision, and innovation across a wide range of applications.

10.3 Case Studies in Thermal and Fluid System Optimization

AI is increasingly being applied to optimize thermal and fluid systems across various industries, leading to significant improvements in energy efficiency, system performance, and cost reduction. By leveraging AI's ability to analyze complex datasets and simulate real-world conditions, industries are developing smarter, more efficient systems that enhance both operational effectiveness and sustainability. Below are key case studies where AI has had a transformative impact.

10.3.1   1. HVAC Systems: Optimizing Energy Efficiency and Comfort

Heating, Ventilation, and Air Conditioning (HVAC) systems are essential for maintaining comfortable indoor environments in homes, offices, and industrial facilities. Traditionally, HVAC systems have been manually controlled or operated based on pre-set schedules, which can lead to inefficiencies, especially in large buildings with varying occupancy levels and external weather conditions. AI has emerged as a game-changer in the optimization of HVAC systems by enabling real-time monitoring and control based on dynamic factors, improving both energy efficiency and occupant comfort.

- **AI-Driven Smart Thermostats and Climate Control:** AI is used in smart thermostats and building management systems to optimize HVAC performance. These systems analyze data from sensors that measure temperature, humidity, and occupancy patterns, and AI algorithms adjust the heating and cooling output in real-time. By predicting when rooms will be occupied or when outdoor temperatures will fluctuate, AI optimizes HVAC operation, reducing energy consumption during off-peak times while maintaining comfort when needed. For example, Google's **Nest Learning Thermostat** uses machine learning to predict user preferences and automatically adjust indoor temperatures, reducing unnecessary heating or cooling.

- **Case Study: Large-Scale Commercial Buildings:** In large commercial buildings, AI-driven HVAC systems have led to substantial energy savings. For instance, a case study from the **Empire State Building** demonstrated that by implementing AI-controlled HVAC systems, the building reduced its energy consumption by over 30%. AI continuously adjusts heating and cooling based on occupancy, external weather conditions, and real-time sensor data, significantly improving energy efficiency while ensuring that occupants remain comfortable throughout the day.

Automotive Cooling Systems: AI for Efficient EV Thermal Management

The rise of electric vehicles (EVs) presents new challenges in cooling systems, particularly when it comes to managing the heat generated by both the battery pack and electric motor. Traditional cooling systems, which are designed for internal combustion engines, are not sufficient for EVs, where maintaining optimal battery temperatures is critical for performance, longevity, and safety. AI plays a key role in optimizing the design and operation of **automotive cooling systems**, helping to ensure that EVs operate efficiently across different driving conditions.

- **AI-Optimized Battery Thermal Management:** AI is used to model and optimize the thermal management of batteries, predicting heat generation patterns based on driving behaviors, external temperatures, and battery load. This is crucial because lithium-ion batteries, which are commonly used in EVs, degrade faster when exposed to high temperatures. AI algorithms can predict the optimal cooling rates, dynamically adjusting cooling system operation to prevent overheating while minimizing energy consumption. Additionally, AI-based predictive models can foresee high-demand scenarios, such as rapid acceleration or extreme weather, and pre-emptively adjust the cooling system to maintain battery efficiency.

- **Case Study: Tesla's AI-Driven Cooling Systems:** Tesla has been a leader in integrating AI into its EV cooling systems. In Tesla's **Model 3**, AI continuously monitors the battery and motor temperatures and adjusts the cooling system based on real-time data. The system optimizes thermal management during high-speed driving, rapid charging, and extreme weather conditions, ensuring that the battery operates within its ideal temperature range. This AI-driven approach extends the battery life, improves performance, and enhances the safety of the vehicle. Additionally, Tesla's **Smart Cooling System** reduces energy consumption by dynamically adjusting the cooling based on environmental and vehicle conditions, resulting in longer driving range.

10.3.2 Industrial Fluid Systems: AI-Driven Flow Optimization

In industries such as oil and gas, chemical processing, and water treatment, fluid transport and processing systems play a vital role in ensuring operational efficiency. Optimizing fluid flow through pipelines, pumps, and valves is essential for minimizing energy consumption, reducing downtime, and improving system reliability. AI is increasingly being deployed to monitor and optimize fluid flow in real-time, leading to improved performance and significant cost savings.

- **AI in Oil and Gas Pipelines:** In the oil and gas industry, pipelines transport massive volumes of fluids over long distances. Maintaining optimal flow conditions is challenging due to friction losses, varying fluid properties, and changes in demand. AI algorithms are applied to monitor the flow rates, pressure levels, and temperature data from sensors along the pipeline. By predicting potential bottlenecks, leaks, or inefficiencies in the system, AI can suggest real-time adjustments to pump speeds, valve positions, and pressure settings, ensuring that the pipeline operates at maximum efficiency.

- **Case Study: Predictive Maintenance in Fluid Systems:** A case study from **Shell** demonstrates how AI has been used to optimize fluid flow in its oil refineries. By implementing AI-based predictive maintenance systems, Shell has been able to reduce downtime and energy consumption in its fluid transport networks. AI analyzes historical data and real-time sensor readings to predict when equipment, such as pumps or compressors, is likely to fail or operate inefficiently. By identifying these issues early, the system can schedule maintenance at optimal times, avoiding costly shutdowns and improving overall system reliability. The predictive maintenance solution also optimizes fluid flow, reducing the energy required to move fluids through the refinery by up to 20%.

- **Water Treatment Systems:** In water treatment plants, AI is used to optimize the flow of water and chemicals during the treatment process. AI models analyze flow rates, pressure levels, and water quality data to predict the optimal operating conditions for pumps and filtration systems. This leads to more efficient water treatment, reducing energy use while improving the quality of the output water.

10.3.3 Future Directions in AI for Thermal and Fluid System Optimization

Looking forward, the role of AI in thermal and fluid system optimization will continue to expand, with new advancements in **AI-driven autonomous systems** and **smart infrastructure**. For instance, AI could enable fully autonomous HVAC systems that not only adjust based on real-time conditions but also anticipate future demand using weather forecasts, occupant schedules, and historical data. In the automotive industry, AI is likely to play a key role in developing next-generation cooling systems for hybrid and fully autonomous vehicles. In industrial fluid systems, AI's ability to process big data and optimize operations will lead to smarter, more sustainable infrastructure in sectors like energy, water management, and manufacturing.

AI's impact on thermal and fluid system optimization is transforming industries by enabling smarter, more efficient systems that respond dynamically to real-world conditions. From optimizing building climate control to improving battery cooling in EVs and enhancing fluid flow in industrial processes, AI is driving innovation and efficiency in ways that were previously unattainable.

11 Chapter 9: Challenges and Ethical Considerations

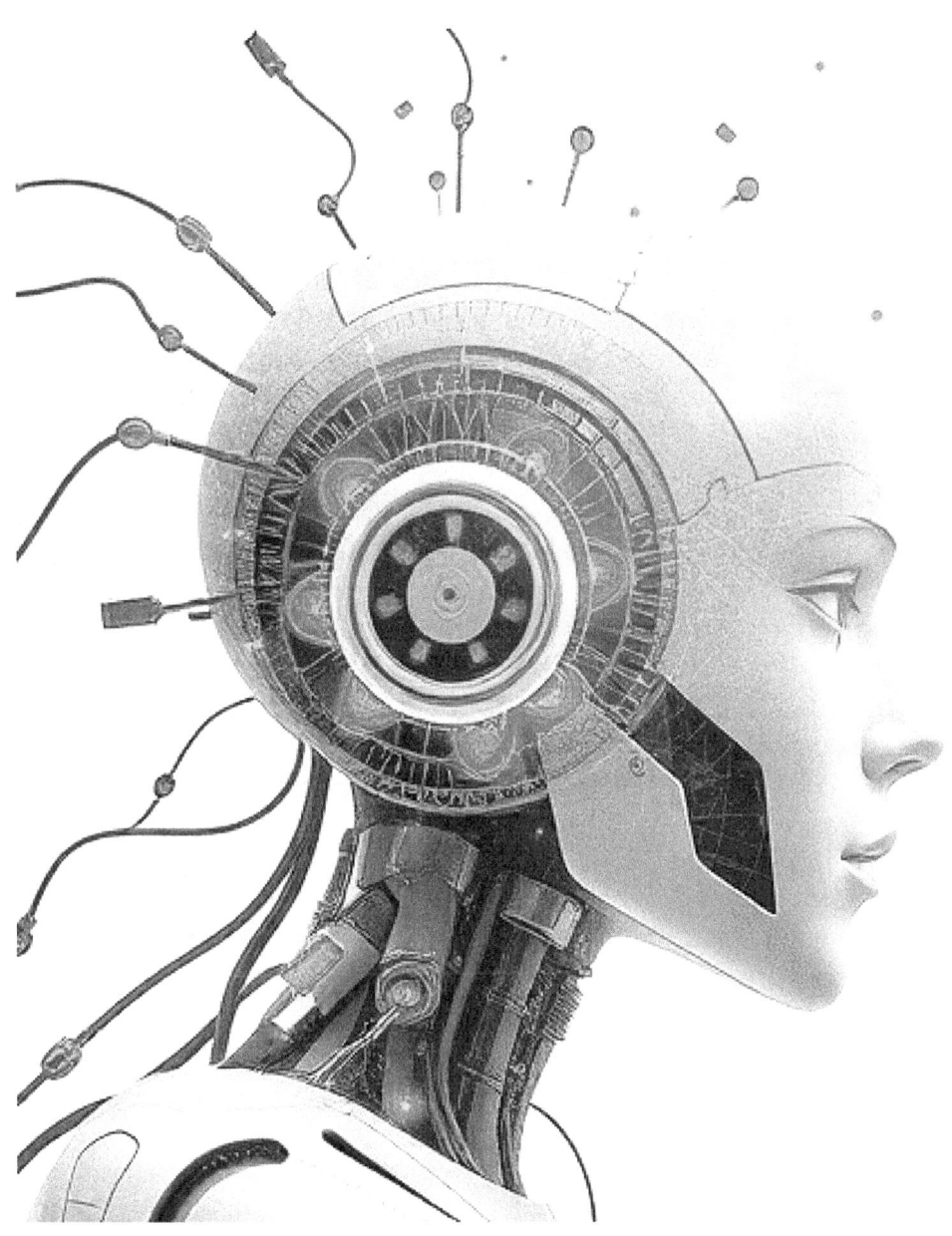

11.1 Challenges in Integrating AI with Mechanical Engineering

The integration of AI into mechanical engineering holds great promise, from optimizing design processes to enhancing predictive maintenance. However, several challenges hinder its seamless application, including:

Data Quality and Availability

One of the most critical challenges in AI integration is ensuring the availability of high-quality data for training AI models. Mechanical systems generate vast amounts of data, but not all of it is structured or useful. Inconsistent data formats, missing values, and noisy data can reduce the accuracy of AI algorithms. Furthermore, in certain applications, data may be scarce, making it difficult to create reliable models. This requires extensive preprocessing, cleaning, and data validation efforts to ensure that the AI models are trained on accurate and representative datasets.

Complexity of AI Algorithms

AI algorithms, particularly in the realms of machine learning and deep learning, are often highly complex. Integrating these algorithms into mechanical systems requires not only understanding AI technology but also the ability to apply it in contexts where physical laws, material science, and system dynamics must be respected. Mechanical systems tend to be governed by deterministic models, while AI relies on probabilistic models, creating a challenge in aligning these two paradigms. Furthermore, real-time implementation can be difficult due to the computational demands of sophisticated AI algorithms.

System Integration

Mechanical engineering systems often rely on established processes, hardware, and software that have been developed and optimized over many years. Integrating AI into these legacy systems without disrupting existing workflows is a significant challenge. Ensuring that AI-driven decisions are reliable and understandable to engineers who manage these systems is critical. Moreover, there is often a need for creating new interfaces that allow AI systems to communicate seamlessly with mechanical control systems, which may not have been designed to handle advanced AI inputs.

Skill Gaps and Interdisciplinary Knowledge

There is a growing demand for professionals who have expertise in both mechanical engineering and AI. While mechanical engineers are experts in fields such as thermodynamics, material science, and kinematics, AI expertise requires a different skill set, including programming, data science, and machine learning. Bridging this skill gap is essential to ensuring the successful integration of AI into mechanical engineering projects.

Educational programs and training need to adapt to this emerging demand for interdisciplinary knowledge.

Ethics and Trust in AI Systems

As AI is applied to mechanical systems that are often responsible for safety-critical tasks, such as automotive engineering or aerospace, concerns about the reliability and ethical use of AI emerge. Engineers need to trust that AI-driven systems will perform safely under all conditions. This requires thorough testing, validation, and regulatory approval to ensure that AI decisions do not compromise system integrity or safety. In addition, transparency in how AI models make decisions is critical for building trust, especially in industries where failure could have catastrophic consequences.

Costs and Resources

The development and deployment of AI systems require significant computational resources, skilled labor, and time for research and development. For small to medium-sized engineering firms, the cost of integrating AI may be prohibitive, particularly when the return on investment is unclear. The need for high-end computing infrastructure to process large datasets and run advanced algorithms can also be a challenge. Additionally, there is the risk of implementation failure if the AI system does not perform as expected, leading to potential financial and operational setbacks.

11.2 Ethical Considerations in AI Applications

As AI continues to permeate various sectors, including mechanical engineering, ethical considerations become increasingly important to ensure that these systems are responsible, fair, and trustworthy. Ethical challenges in AI applications include:

Bias and Fairness

AI models learn from data, and if that data contains inherent biases, the AI can replicate or even amplify those biases. This is a significant ethical concern, especially in applications that affect human decision-making or public safety, such as autonomous vehicles or industrial automation systems. In these cases, biased AI could lead to unequal outcomes, safety risks, or discrimination against certain groups.

For example, an AI system used in manufacturing quality control could be trained on data that inadvertently favors one material type over another, leading to faulty decisions. It is essential to implement strategies that detect and mitigate biases, such as diverse data sourcing, algorithm audits, and fairness checks, to ensure AI systems operate equitably.

Transparency and Explainability

Many AI systems, particularly deep learning models, function as "black boxes," where it is difficult to understand how they arrive at specific decisions. In critical applications, such as predictive maintenance of mechanical systems or autonomous robotics, transparency is crucial. Engineers and operators need to understand the decision-making process of AI systems to trust them and troubleshoot errors when they arise.

For instance, if an AI system controlling an autonomous vehicle or an industrial robot makes an unexpected decision, engineers must be able to trace the logic behind that decision. To address this, explainable AI (XAI) is becoming an important field, focusing on making AI decision-making processes more interpretable without compromising performance.

Privacy and Data Protection

AI often relies on vast amounts of data, some of which may include sensitive or personal information. In areas like predictive maintenance, smart manufacturing, or AI-driven diagnostics, privacy concerns can arise when personal or proprietary data is collected. Ethical AI applications require strict adherence to data privacy laws, such as GDPR (General Data Protection Regulation) in Europe or CCPA (California Consumer Privacy Act) in the U.S., which aim to protect user data and give individuals more control over their information.

In mechanical engineering, while personal data may not be as prevalent, proprietary data such as design specifications, operational performance, or manufacturing processes may still be sensitive. Protecting this data from unauthorized access and ensuring it is used responsibly is a critical ethical concern. AI developers must implement secure data practices, including anonymization, encryption, and access control, to mitigate the risk of data breaches or misuse.

Accountability and Liability

When AI systems fail or make erroneous decisions, the question of accountability arises. Determining who is responsible for AI-driven decisions is critical, particularly in high-stakes applications such as aerospace, automotive engineering, or industrial safety. If an AI system malfunctions, causing damage to machinery or even injury to humans, there must be clear legal frameworks to determine liability—whether it lies with the AI developer, the data provider, or the operator of the system.

Ethical AI practices demand that organizations using AI clearly define responsibility for AI outcomes. This includes creating protocols for human oversight, establishing clear documentation of AI decision-making processes, and ensuring that AI systems are robustly tested in real-world scenarios before being deployed.

Safety and Reliability

In mechanical engineering applications where AI interacts with physical systems, safety becomes a paramount ethical consideration. AI systems controlling autonomous vehicles, robotics, or industrial equipment must function reliably in all conditions, including unpredictable or extreme environments. Failure to do so could lead to accidents, equipment damage, or even loss of life.

Ensuring the safety and reliability of AI systems involves rigorous testing, validation, and continuous monitoring. This also means developing AI models that can operate with a degree of fail-safes and redundancies in place, especially in mission-critical environments.

11.3 Future Directions and Innovations

As AI continues to transform mechanical engineering, future advancements will rely on key strategies to foster innovation and ensure the responsible integration of AI into engineering practices. Some critical future directions include:

Interdisciplinary Collaboration

One of the most important drivers of AI innovation in mechanical engineering will be the promotion of interdisciplinary collaboration. As AI becomes more integrated into traditional engineering domains, it is essential for AI experts, data scientists, and mechanical engineers to work together closely. Each discipline brings unique strengths—AI professionals offer expertise in data analysis, algorithms, and machine learning, while mechanical engineers provide deep knowledge of physical systems, materials, and design principles.

Successful future innovations will depend on teams that bridge these two worlds. For example, in robotics, AI can optimize motion planning or predict mechanical wear, but the integration of these AI systems requires engineers who understand both mechanical structures and AI processes. Universities, research institutions, and industries are already fostering interdisciplinary education, but future trends will likely focus on creating cross-functional teams and integrated research labs where both AI and mechanical engineering experts collaborate from the earliest stages of project development.

Development of Regulations and Standards

As AI becomes more prevalent in critical engineering applications, there is a growing need for the development of regulations and standards that ensure the ethical, safe, and reliable use of AI in engineering fields. Current regulatory frameworks often lag behind technological advancements, which can leave gaps in safety, accountability, and ethical considerations. As a result, governing bodies and international organizations need to work on creating robust guidelines tailored to the use of AI in fields like mechanical engineering, aerospace, automotive systems, and manufacturing.

These regulations will likely cover several areas, including:

- **Safety standards** for AI-driven machinery and autonomous systems.
- **Data protection regulations** that safeguard the use of proprietary or sensitive data in AI models.
- **Ethical guidelines** to ensure fairness, transparency, and accountability in AI decision-making processes.
- **Testing and validation protocols** to ensure AI models perform reliably in all expected operating conditions.

Moreover, international cooperation will be essential to create unified global standards that facilitate the safe deployment of AI technologies across industries and borders.

Continuous Learning and Adaptation

AI is a rapidly evolving field, with new algorithms, tools, and techniques emerging constantly. For AI to remain valuable in mechanical engineering, professionals in both fields must commit to continuous learning and adaptation. AI algorithms, once deployed, should be designed to evolve over time as new data becomes available. This is particularly relevant in predictive maintenance, smart manufacturing, and autonomous systems, where real-world conditions can vary widely.

Mechanical engineers will also need ongoing training in AI and data science to keep pace with technological advancements. This may require new educational pathways that focus on lifelong learning, incorporating AI-related courses into engineering curricula, as well as offering industry-based workshops, certifications, and online learning platforms. Similarly, AI specialists will need to stay informed about advancements in mechanical engineering to ensure their AI solutions are effective in practical, real-world settings.

AI-Enhanced Design and Simulation

One of the most promising future innovations lies in the area of AI-enhanced design and simulation. AI has already shown its potential in optimizing designs for performance, weight, material use, and cost. In the future, AI could take this further by creating autonomous design systems capable of generating optimized mechanical designs based on specific criteria, significantly reducing design time and costs.

AI-driven simulations will also grow more sophisticated, allowing engineers to test complex mechanical systems under a wide range of scenarios, including those that would be difficult or impossible to simulate in the physical world. This capability will be especially valuable in fields like aerospace, automotive engineering, and renewable energy, where AI models can simulate long-term system performance, fatigue analysis, or the behavior of new materials under stress.

Sustainable Engineering and AI

The future of AI in mechanical engineering also intersects with sustainability goals. AI technologies can be applied to optimize energy consumption, reduce waste in manufacturing processes, and improve the efficiency of renewable energy systems. For instance, AI could be used to design more efficient wind turbines, optimize fuel consumption in transportation systems, or reduce material waste in 3D printing processes. As environmental concerns continue to grow, AI will play a critical role in helping mechanical engineers design more sustainable systems and practices.

Human-AI Collaboration

In the future, the relationship between human engineers and AI will evolve into more collaborative dynamics, where AI systems act as co-pilots rather than autonomous decision-makers. Human engineers will leverage AI's analytical power while maintaining oversight and control over decision-making processes. This hybrid approach allows engineers to focus on high-level problem-solving and creative tasks while AI handles repetitive, data-heavy computations.

12 Chapter 10: The Future of AI in Mechanical Engineering

12.1 Emerging Trends and Technologies

12.1.1 Emerging Trends and Technologies

The intersection of AI and mechanical engineering is giving rise to several emerging trends and technologies that are poised to revolutionize how engineers design, analyze, and maintain systems. These technologies are pushing the boundaries of what is possible in engineering, from real-time monitoring to advanced simulations. Key emerging trends include:

12.1.1.1 AI-Enhanced Simulation

AI is transforming the world of simulations in mechanical engineering by improving both the accuracy and speed of these processes. Traditional simulations, such as finite element analysis (FEA) or computational fluid dynamics (CFD), can be computationally expensive and time-consuming. AI, especially machine learning algorithms, is being used to accelerate these simulations by learning from previous simulation data and predicting outcomes in new scenarios much faster.

AI-enhanced simulations are particularly beneficial in areas like materials science, where AI can predict how new materials will behave under stress or extreme conditions without the need for extensive physical testing. AI can also optimize simulations by identifying the most critical variables, reducing the overall complexity of the model without sacrificing accuracy.

By integrating AI into simulation workflows, mechanical engineers can explore a greater number of design iterations in less time, enabling faster innovation in product development, particularly in industries such as aerospace, automotive, and energy.

12.1.1.2 Digital Twins

A digital twin is a virtual model of a physical system that can simulate its performance in real-time. These digital replicas are powered by AI and IoT (Internet of Things) technologies, and they allow engineers to monitor, analyze, and optimize physical systems more efficiently than ever before. Digital twins are being used in a variety of applications, from smart manufacturing plants to aircraft engines, to monitor real-time performance, predict maintenance needs, and optimize operations.

The AI-driven aspect of digital twins allows for predictive analytics, meaning that potential failures or performance inefficiencies can be identified before they occur. This can lead to substantial cost savings by preventing unplanned downtime and improving the lifespan of machinery. For example, in aerospace, a digital twin of an aircraft engine can analyze real-time data from sensors on the engine to predict maintenance needs, improving safety and operational efficiency.

Digital twins also offer the potential for simulation-based training, where engineers can practice scenarios or test modifications in a digital environment before implementing them in the physical world.

12.1.1.3 Augmented Reality (AR) Combined with AI

The combination of AR and AI is creating powerful tools for engineers, particularly in areas such as design visualization, assembly processes, and maintenance tasks. AR overlays digital information onto the physical world, allowing engineers and technicians to interact with digital models in real-time. When combined with AI, AR systems can provide smart, context-aware information that enhances decision-making and troubleshooting.

For example, in manufacturing, AR combined with AI can be used to guide workers through complex assembly tasks by displaying step-by-step instructions directly on the physical parts they are working with. AI algorithms can monitor the process in real-time, identifying errors or inefficiencies and suggesting corrective actions. In maintenance, AR glasses equipped with AI can overlay diagnostic information directly onto equipment, helping engineers identify issues quickly and accurately without referring to manuals or external systems.

AR also enhances design collaboration, allowing engineers to visualize 3D models of new products in real-world environments before they are physically built. This speeds up the design review process and helps teams identify potential issues early, leading to more efficient product development cycles.

12.1.1.4 Generative Design

Generative design is an AI-driven technology that automatically generates optimized design solutions based on specific constraints and objectives. Engineers input design goals—such as weight reduction, material efficiency, or structural strength—and AI algorithms explore thousands of potential design configurations, selecting those that best meet the requirements. This approach allows for more innovative and unconventional designs that might not be discovered through traditional engineering methods.

Generative design is becoming increasingly popular in industries like automotive and aerospace, where lightweight materials and complex geometries are critical for performance. AI's ability to evaluate a vast design space quickly makes generative design a valuable tool for reducing material use, improving performance, and cutting development time.

12.1.1.5 Robotics and Autonomous Systems

Robotics, already a key area in mechanical engineering, is being further enhanced by AI. Autonomous robots capable of learning and adapting to their environments are increasingly used in manufacturing, logistics, and inspection processes. AI-powered robots can perform complex tasks such as quality control inspections, precision assembly, and even hazardous maintenance operations in environments too dangerous for humans.

AI-driven robotics systems use machine learning algorithms to improve their capabilities over time, making them more flexible and efficient. For example, in the automotive industry, AI-powered robots are used to weld, assemble, and inspect vehicles with a level of precision and consistency that improves product quality and reduces production costs. Autonomous vehicles, both in manufacturing plants and on public roads, represent the next step in AI-robotics integration.

12.1.1.6 Edge AI for Real-Time Decision Making

Edge AI refers to the deployment of AI algorithms directly on devices or local servers, enabling real-time data processing without the need to send data to the cloud. In mechanical engineering, edge AI can be particularly useful for applications that require low latency, such as real-time control systems in robotics or autonomous vehicles, predictive maintenance systems, and smart manufacturing processes.

By processing data locally, edge AI can provide immediate insights and decisions, making it suitable for time-sensitive applications. For example, in industrial equipment, edge AI can detect anomalies in machine performance as they happen, triggering automatic adjustments or alerts to prevent breakdowns. This real-time capability will be critical as more mechanical systems are automated and need to operate independently.

12.2 The Road Ahead: AI in Engineering Education and Research

12.2.1 The Road Ahead: AI in Engineering Education and Research

As AI becomes more integrated into engineering, its influence on education and research is growing rapidly. Preparing the next generation of engineers to leverage AI's capabilities will be crucial for driving innovation and solving complex global challenges. Key areas of focus for AI in engineering education and research include:

12.2.1.1 1. Curriculum Development: Incorporating AI and Machine Learning into Engineering Curricula

One of the most pressing needs in engineering education is the integration of AI and machine learning into traditional engineering curricula. Historically, mechanical engineering programs have focused on core subjects such as thermodynamics, material science, and structural analysis. However, with the rise of AI, there is a growing need for engineers to understand algorithms, data analysis, and computational modeling.

Future engineering programs will need to balance these traditional subjects with AI-focused coursework. Topics such as machine learning, neural networks, data science, and AI ethics will become essential parts of the curriculum. Additionally, hands-on projects and labs where students can apply AI to solve real-world engineering problems will help bridge the gap between theory and practice.

Universities and engineering schools are beginning to offer interdisciplinary programs that combine AI with mechanical engineering, but more widespread adoption will be necessary to meet the increasing demand for AI-skilled engineers. These programs will also need to focus on cultivating a strong foundation in coding, statistics, and data manipulation, as these skills are critical for effective AI applications in engineering.

12.2.1.2 2. Research Collaboration: Encouraging Interdisciplinary Research Collaborations

AI's true potential in mechanical engineering will be realized through interdisciplinary research that combines the expertise of AI specialists with engineers. Research collaborations between AI experts, mechanical engineers, materials scientists, and data analysts will drive new innovations in areas like robotics, autonomous systems, and smart manufacturing.

For example, in the field of energy, collaborations between AI researchers and engineers could lead to breakthroughs in renewable energy systems, optimizing the efficiency of solar panels or improving the design of wind turbines through AI-driven simulations. In healthcare engineering, AI and mechanical engineering researchers could collaborate to develop more advanced medical devices, such as AI-powered prosthetics or robotics-assisted surgery tools.

To foster these collaborations, research institutions and universities must create platforms that encourage cross-disciplinary work, such as joint research centers, conferences, and collaborative funding opportunities. These initiatives will be essential for solving complex, real-world problems that span multiple engineering domains.

12.2.1.3 3. Lifelong Learning: Promoting Lifelong Learning Opportunities for Engineers

AI is a rapidly evolving field, and for engineers to stay competitive, they will need to engage in continuous learning throughout their careers. Lifelong learning opportunities will be critical as new AI technologies, tools, and methodologies continue to emerge. Engineers, regardless of their industry or specialty, will need access to ongoing education to stay up-to-date with advancements in AI.

This could be achieved through several channels:

- **Professional development programs**: Offering AI-focused certifications, workshops, and short courses aimed at working professionals in engineering fields. These programs can provide practical, hands-on experience with AI applications, enabling engineers to apply AI directly to their projects.

- **Online learning platforms**: Massive Open Online Courses (MOOCs), online bootcamps, and webinars that focus on AI and machine learning for engineers. These platforms make it easier for professionals to access learning resources at their own pace and stay informed about the latest AI techniques and tools.

- **Industry partnerships**: Collaboration between academic institutions and industries to create tailored AI training programs that meet specific industry needs. For instance, an aerospace company might partner with a university to develop a specialized AI training program for its engineers, focusing on AI in flight dynamics or predictive maintenance.

Lifelong learning is not just about keeping pace with AI advancements but also about adapting to new challenges. As AI technologies continue to disrupt traditional engineering practices, engineers will need to continually expand their skill sets to remain innovative and competitive in a rapidly changing field.

12.2.1.4 *4. AI in Engineering Research: Expanding Research Frontiers*

AI is opening new frontiers in engineering research. Researchers are now exploring areas that were previously difficult to study due to computational or physical limitations. AI allows for the analysis of vast amounts of data, the modeling of complex systems, and the automation of experimentation, making it possible to tackle more ambitious research goals.

In mechanical engineering, AI-driven research is helping engineers simulate complex physical phenomena such as fluid dynamics, structural integrity, and thermal behavior more accurately and efficiently. AI can also automate repetitive research tasks, freeing up researchers to focus on high-level problem-solving and innovation.

Furthermore, AI is transforming the research process itself by enabling researchers to quickly sift through large datasets, identify patterns, and generate new hypotheses. This accelerates the pace of discovery and innovation in fields ranging from materials science to robotics.

13 Conclusion

The integration of AI into mechanical engineering marks a profound transformation in how we design, manufacture, and maintain the physical systems that shape our world. This fusion of disciplines—where cutting-edge algorithms meet the principles of mechanical engineering—has the potential to revolutionize every facet of the field, from product development to operational efficiency, safety, and sustainability.

As we stand on the cusp of this new era, it's clear that the impact of AI will be far-reaching. AI-driven tools are already enhancing simulations, enabling engineers to model complex systems with greater speed and accuracy, pushing the boundaries of what we can design and build. Digital twins—digital replicas of physical systems—allow for real-time monitoring, predictive maintenance, and performance optimization, drastically reducing downtime and operational costs. The advent of generative design, powered by AI, is unlocking unprecedented opportunities for innovation, producing designs that are lighter, stronger, and more efficient than ever before.

However, alongside these remarkable opportunities, the integration of AI into mechanical engineering also presents significant challenges. Engineers must grapple with the quality of data that fuels AI systems. Inaccurate, incomplete, or biased data can lead to suboptimal or even unsafe AI-driven decisions. As AI models become more complex, ensuring that they integrate seamlessly with existing systems—many of which have been optimized over decades of traditional engineering practice—poses a considerable technical hurdle. The skills gap is another pressing issue, as today's engineers must not only master traditional mechanical engineering but also acquire deep expertise in AI, machine learning, and data analytics.

Moreover, the ethical considerations that accompany the deployment of AI cannot be overlooked. As AI becomes more autonomous and embedded in safety-critical systems like autonomous vehicles, industrial robots, and infrastructure monitoring, ensuring fairness, transparency, and accountability in AI decision-making becomes paramount. AI systems must be explainable, allowing engineers to understand, trust, and, when necessary, override the decisions made by algorithms. Privacy concerns, particularly when sensitive operational or personal data is involved, must also be addressed through robust data protection measures.

To successfully navigate these challenges, the field of mechanical engineering must undergo a paradigm shift. Education will play a central role in preparing the next generation of engineers for this AI-driven future. Engineering curricula must evolve to incorporate AI, data science, and machine learning alongside core mechanical engineering subjects. Universities and industry alike need to foster interdisciplinary collaboration, ensuring that mechanical engineers can work seamlessly with AI specialists, computer scientists, and data analysts to solve complex problems. Research institutions must lead the charge, exploring new applications of AI in areas such as renewable energy, advanced manufacturing, healthcare technologies, and aerospace.

Lifelong learning will also be crucial. As AI continues to evolve at a rapid pace, engineers will need to continuously update their skills and knowledge to remain competitive and innovative. Professional development programs, online courses, and industry certifications will help ensure that mechanical engineers are equipped with the latest AI tools and techniques. Continuous education will enable engineers not only to adapt to new AI technologies but also to apply them creatively to solve emerging challenges in areas such as climate change, sustainable manufacturing, and smart cities.

Looking ahead, the integration of AI into mechanical engineering holds immense promise for building a more efficient, safe, and sustainable future. AI's potential to optimize processes, reduce waste, and improve energy efficiency aligns perfectly with the growing global demand for sustainability. AI-enabled smart manufacturing, for instance, can reduce material waste, energy consumption, and environmental impact, while autonomous systems can help engineers design and operate renewable energy systems more efficiently.

Furthermore, AI is driving the development of intelligent systems that can autonomously learn, adapt, and improve over time. In mechanical engineering, this could mean everything from self-repairing machines to autonomous vehicles that operate safely and efficiently in unpredictable environments. The combination of AI and advanced robotics promises to redefine how we approach tasks that are dangerous, repetitive, or require precision beyond human capability.

In conclusion, the road ahead for AI in mechanical engineering is both exciting and challenging. The full realization of AI's transformative potential will require careful planning, ongoing education, interdisciplinary collaboration, and a commitment to ethical principles. As engineers, educators, researchers, and policymakers work together to address these challenges, the future of mechanical engineering will be defined not just by the power of AI, but by our ability to harness it responsibly and effectively.

By integrating AI thoughtfully, we can ensure that this technology becomes a powerful tool for solving the complex problems of tomorrow—paving the way for innovations that will not only enhance the capabilities of mechanical engineering but also contribute to the creation of a smarter, safer, and more sustainable world.

14 Appendices

Kaps, I., et al. (2020). "AI in water treatment: A comprehensive review." *Environmental Science & Technology*

Kermanshah, A., et al. (2019). "Predictive maintenance for oil and gas pipelines using artificial intelligence." *Journal of Petroleum Science and Engineering*

Wang, Y., et al. (2021). "A comprehensive review of thermal management systems for electric vehicles." *Applied Thermal Engineering*

Chen, H., et al. (2020). "Application of artificial intelligence in HVAC systems

Ferrari, A., & Willcox, K. (2024). "Digital Twins in Mechanical and Aerospace Engineering." *Nature Computational Science*, 4, 178–183. https://doi.org/10.1038/s43588-024-00613-8.

Karniadakis, G. E., Kevrekidis, I. G., Lu, L., Perdikaris, P., Wang, S., & Yang, L. (2021). "Physics-Informed Machine Learning." *Nature Reviews Physics*, 3, 422–440. https://doi.org/10.1038/s43588-021-00217-1.

Peherstorfer, B., & Willcox, K. (2016). "Data-Driven Operator Inference for Nonintrusive Projection-Based Model Reduction." *Computer Methods in Applied Mechanics and Engineering*, 306, 196–215. https://doi.org/10.1016/j.cma.2016.03.025.

Brunton, S. L., Proctor, J. L., & Kutz, J. N. (2016). "Discovering Governing Equations from Data by Sparse Identification of Nonlinear Dynamical Systems." *Proceedings of the National Academy of Sciences*, 113(15), 3932–3937. https://doi.org/10.1073/pnas.1517384113.

Baker, N., Alexander, F., Bremer, T., Hagberg, A., Kevrekidis, I., Najm, H., Parashar, M., Patra, A., Sethian, J., Wild, S., & Willcox, K. (2019). "Workshop Report on Basic Research Needs for Scientific Machine Learning: Core Technologies for Artificial Intelligence." *U.S. DOE Office of Science*. https://doi.org/10.2172/1484330.

Zhou, Y., Wang, M., & Chen, L. (2021). "Artificial Intelligence in Fluid Mechanics: A Review." *Annual Review of Fluid Mechanics*, 53, 1–30. https://doi.org/10.1146/annurev-fluid-060220-113730.

Willcox, K., & Ghattas, O. (2021). "Learning Physics-Based Models from Data: Perspectives from Inverse Problems and Model Reduction." *Acta Numerica*, 30, 445–554. https://doi.org/10.1017/S0962492920000151.

Jafary, T., Li, Z., & Liu, Y. (2023). "Ethical Implications of Artificial Intelligence in Mechanical Engineering." *Journal of Mechanical Engineering and Ethics*, 40(2), 123–139. https://doi.org/10.1016/j.jme.2023.01.006.

Sanderson, C., & Pati, Y. C. (2023). "AI in Engineering Design and Optimization: A Systematic Review." *International Journal of Mechanical Engineering*, 58(5), 221–237. https://doi.org/10.1177/09544062211001435.

www.ingramcontent.com/pod-product-compliance
Lightning Source LLC
Chambersburg PA
CBHW062121220526

45471CB00010B/3832